PACIFIC PROFILES

VOLUME SEVEN
Allied Transports: Douglas C-47 Series
South & Southwest Pacific 1942–1945

MICHAEL JOHN CLARINGBOULD

Avonmore Books

Pacific Profiles Volume Seven

Allied Transports: Douglas C-47 Series South & Southwest Pacific 1942–1945

Michael John Claringbould

ISBN: 978-0-6452469-1-9

First published 2022 by Avonmore Books

Avonmore Books
PO Box 217
Kent Town
South Australia 5071
Australia

Phone: (61 8) 8431 9780
avonmorebooks.com.au

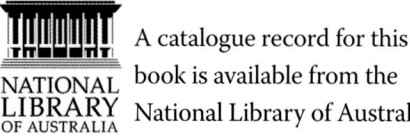 A catalogue record for this book is available from the National Library of Australia

Cover design & layout by Diane Bricknell

Front Cover: A suite of different markings and camouflage schemes attended Douglas transport operations in both the SWPA and SOPAC theatres (top to bottom): a former NEIAF DC-3 later assigned to the 374th TCG as VHCXE (Profile 70), The Wolf (not profiled) from the 13th TCS, NEIAF C-47 Grobak (Profile 98), Douglas C-50 in natural metal finish as operated by Guinea Airways in 1944 (Profile 75) and RAAF C-47 A65-114 (Profile 93).

Rear Cover: Douglas C-50 serial 41-7698 was reassigned to Guinea Airways in May 1944. Here it approaches Wards 'drome from Townsville against the backdrop of the Owen Stanley ranges.

Contents

About the Author 5

Glossary and Abbreviations 7

Introduction 9

Chapter 1 – Overview of C-47 Operations in the Southwest and South Pacific 11

Chapter 2 – C-47 Markings in the SOPAC and SWPA 17

Chapter 3 – 6th Troop Carrier Squadron "Bully Beef Express" ...23

Chapter 4 – 13th Troop Carrier Squadron "The Thirsty 13th" 27

Chapter 5 – 21st and 22nd Troop Carrier Squadrons 35

Chapter 6 – 33rd Troop Carrier Squadron 41

Chapter 7 – 39th Troop Carrier Squadron 47

Chapter 8 – 40th Troop Carrier Squadron 51

Chapter 9 – 41st Troop Carrier Squadron 57

Chapter 10 – 46th Troop Carrier Squadron 65

Chapter 11 – 55th Troop Carrier Squadron 71

Chapter 12 – 56th Troop Carrier Squadron 75

Chapter 13 – 57th Troop Carrier Squadron "King's Men" 79

Chapter 14 – 58th Troop Carrier Squadron 83

Chapter 15 – 63rd Troop Carrier Squadron 87

Chapter 16 – 64th Troop Carrier Squadron 89

Chapter 17 – 65th Troop Carrier Squadron 93

Chapter 18 – 66th Troop Carrier Squadron 97

Chapter 19 – 67th Troop Carrier Squadron 103

Chapter 20 – 68th Troop Carrier Squadron 107

Chapter 21 – 69th Troop Carrier Squadron 111

Chapter 22 – 70th Troop Carrier Squadron 115

Chapter 23 – Directorate of Air Transport 119

Chapter 24 – Miscellaneous C-47s 131

Chapter 25 – United States Marine Corps 135

Chapter 26 – Royal Australian Air Force 143

Chapter 27 – Royal New Zealand Air Force 149

Chapter 28 – Dutch Transport Units 153

Sources & Acknowledgements 158

Dedication 159

Index of Names 160

The author with Papua New Guinea friends in Port Moresby in 2004.

About the Author

Michael Claringbould – Author & Illustrator

Michael spent his formative years in Papua New Guinea in the 1960s, during which he became fascinated by the many WWII aircraft wrecks which lay around the country and also throughout the Solomon Islands. Michael subsequently served widely overseas as an Australian diplomat throughout Southeast Asia and the Pacific, including Fiji (1995-1998) and Papua New Guinea (2003-2005). Michael has authored and illustrated numerous books on Pacific War aviation. His history of the Tainan Naval Air Group in New Guinea, *Eagles of the Southern Sky*, received worldwide acclaim as the first English-language history of a Japanese fighter unit, and was subsequently translated into Japanese. An executive member of Pacific Air War History Associates, Michael holds a pilot license and PG4 paraglider rating. He continues to develop his skills as a digital aviation artist and cartoonist.

Other Books by the Author

Black Sunday (2000)

Eagles of the Southern Sky (2012, with Luca Ruffato)

F4U Corsair versus A6M2/3/4 Zero-sen, Solomons and Rabaul 1943-44 (Osprey, 2022)

Nemoto's Travels The illustrated saga of a Japanese floatplane pilot in the first year of the Pacific War (2021)

Operation I-Go Yamamoto's Last Offensive – New Guinea and the Solomons April 1943 (2020)

P-39 / P-400 Airacobra versus A6M2/3 Zero-sen New Guinea 1942 (Osprey, 2018)

P-47D Thunderbolt versus Ki-43 Hayabusa New Guinea 1943/44 (Osprey, 2020)

Pacific Adversaries Volume One Japanese Army Air Force vs The Allies New Guinea 1942-1944 (2019)

Pacific Adversaries Volume Two Imperial Japanese Navy vs The Allies New Guinea & the Solomons 1942-1944 (2020)

Pacific Adversaries Volume Three Imperial Japanese Navy vs The Allies New Guinea & the Solomons 1942-1944 (2020)

Pacific Adversaries Volume Four Imperial Japanese Navy vs The Allies - The Solomons 1943-1944 (2021)

Pacific Profiles Volume One Japanese Army Fighters New Guinea & the Solomons 1942-1944 (2020)

Pacific Profiles Volume Two Japanese Army Bomber & Other Units, New Guinea and the Solomons 1942-44 (2020)

Pacific Profiles Volume Three Allied Medium Bombers, A20 Series, South West Pacific 1942-44 (2020)

Pacific Profiles Volume Four Allied Fighters: Vought F4U Corsair Series Solomons Theatre 1943-1944 (2021)

Pacific Profiles Volume Five Japanese Navy Zero Fighters (land-based) New Guinea and the Solomons 1942-1944 (2021)

Pacific Profiles Volume Six Allied Fighters: Bell P-39 & P-400 Airacobra South & Southwest Pacific 1942-1944 (2022)

Pacific Profiles Volume Eight: IJN Floatplanes in the South Pacific 1942-1944 (2022)

South Pacific Air War Volume 1: The Fall of Rabaul December 1941–March 1942 (2017, with Peter Ingman)

South Pacific Air War Volume 2: The Struggle for Moresby March–April 1942 (2018, with Peter Ingman)

South Pacific Air War Volume 3: Coral Sea & Aftermath May-June 1942 (2019, with Peter Ingman)

South Pacific Air War Volume 4: Buna & Milne Bay June-September 1942 (2020, with Peter Ingman)

South Pacific Air War Volume 5: Crisis in Papua September – December 1942 (2022, with Peter Ingman)

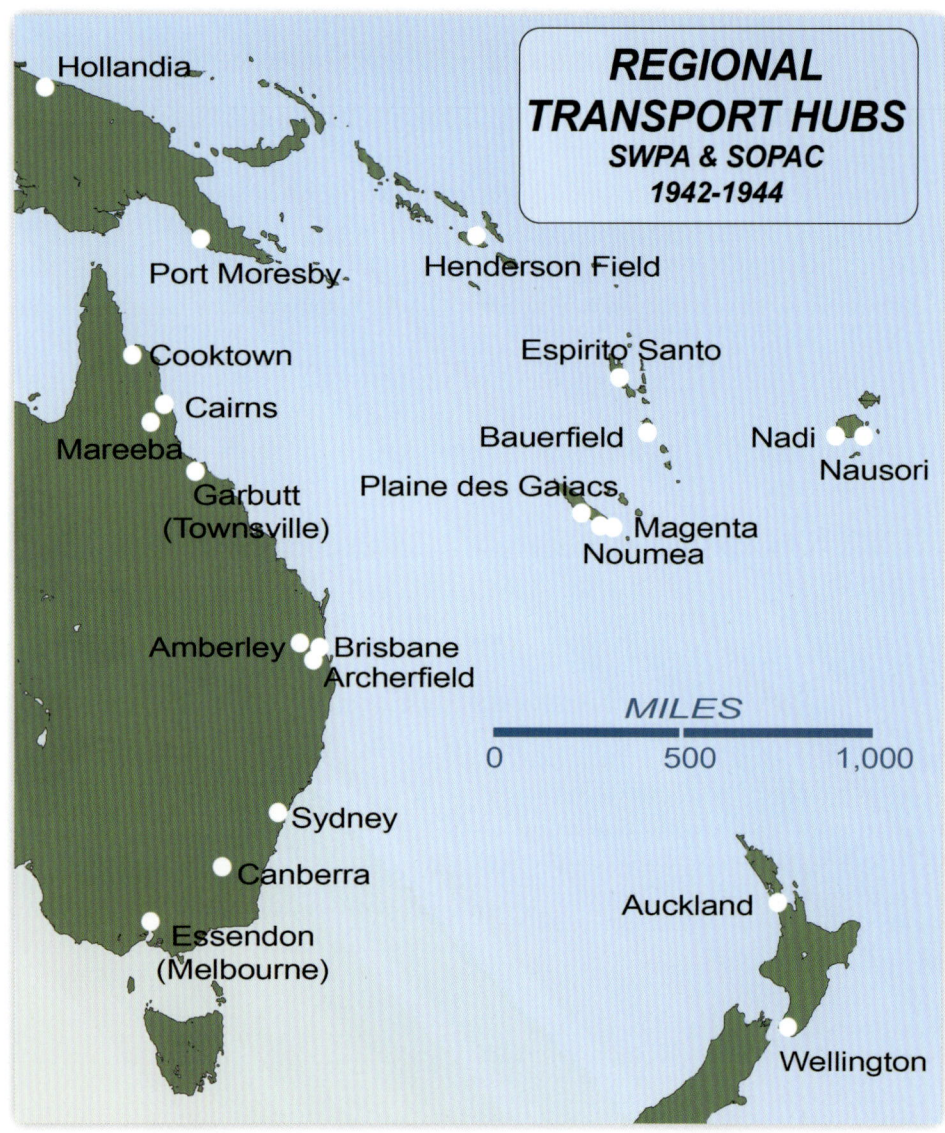

REGIONAL TRANSPORT HUBS
SWPA & SOPAC
1942-1944

Hollandia

Port Moresby

Henderson Field

Cooktown

Espirito Santo

Cairns

Mareeba

Bauerfield

Nadi

Garbutt
(Townsville)

Plaine des Gaiacs

Nausori

Magenta

Noumea

Amberley

Brisbane

Archerfield

MILES

0   500   1,000

Sydney

Canberra

Essendon
(Melbourne)

Auckland

Wellington

This map shows many of the airfields that were commonly used throughout the South West Pacific Area (SWPA) and the South Pacific (SOPAC) theatre. Broadly, everything to the west of Australia and New Guinea was in the SOPAC, while Australia and New Guinea made up the SWPA. While not shown on this map, there was much aerial transport activity to locations in the west and north of Australia also.

Glossary and Abbreviations

ATC	Air Transport Command
AWOL	Absent without leave
BuAer	Bureau of Aeronautics; this was the USN's material support organisation for naval aviation and allocated serial numbers to all USN and USMC aircraft.
DAT	Directorate of Air Transport
FEAF	Far East Air Force
IJA	Imperial Japanese Army
Kokutai	Japanese naval air group
MAG	Marine Aircraft Group
NEIAF	Netherlands East Indies Air Force
POW	Prisoner of War
RAAF	Royal Australian Air Force
RAF	Royal Air Force
RNZAF	Royal New Zealand Air Force
SCAT	South Pacific Combat Air Transport Command
Sentai	Abbreviation of *hiko sentai* defining a Japanese Army Air Force flying regiment.
SMS-	Service Marine Squadron
SOPAC	South Pacific Area
SWPA	South West Pacific Area
TA	Transport *Afdeeling* (Transport Section or Flight, Dutch)
TCG	Troop Carrier Group
TCS	Troop Carrier Squadron
TCW	Troop Carrier Wing
TS	Transport Squadron
US	United States
USAAF	United States Army Air Force
USMC	United States Marine Corps
USN	United States Navy
VMJ-	Prefix for USMC utility squadron
VMR-	Prefix for USMC transport squadron

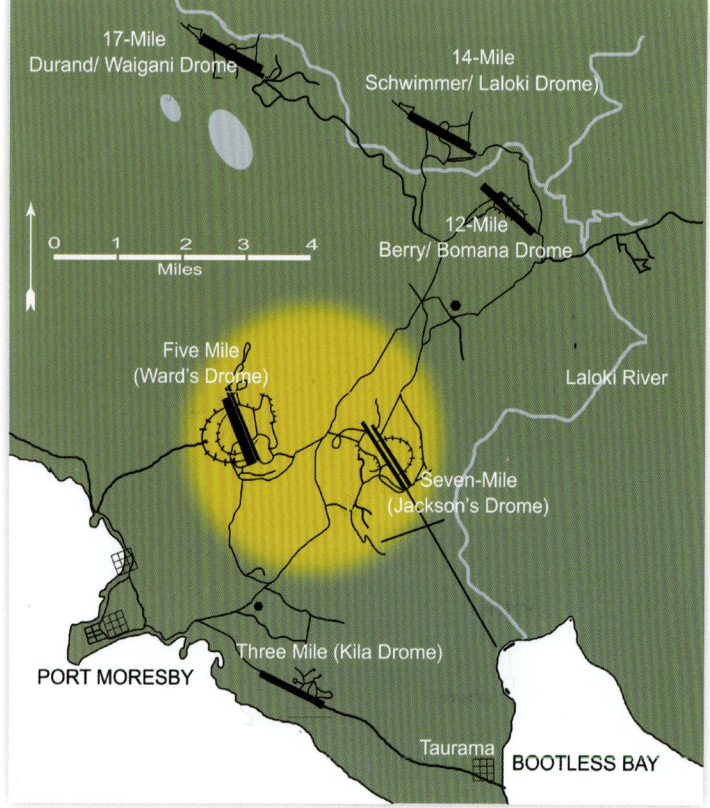

FRONTLINE AIRFIELDS
SWPA & SOPAC Theatres
1942-1944

Momote

Tadji

Madang

Garoka

Mt Hagen

Dumpu

Gusap

Kaiapit

Nadzab

Tsile Tsile

Lae

Wau

Cape Gloucester

Saidor

Finschhafen

Dobodura

Kokoda

Kiriwina

Terapo

Port Moresby

Milne Bay

Torokina

Stirling

Barakoma

Munda

Ondonga

Segi

Banika

Yandina

Henderson Field

Many of the frontline airfields referred to in this volume in New Guinea and the Solomons.

17-Mile
Durand/ Waigani Drome)

14-Mile
Schwimmer/ Laloki Drome)

12-Mile
Berry/ Bomana Drome)

Five Mile
(Ward's Drome)

Laloki River

Seven-Mile
(Jackson's Drome)

Three Mile (Kila Drome)

PORT MORESBY

Taurama

BOOTLESS BAY

0 1 2 3 4
Miles

The wider Port Moresby area where during the course of 1942 four new 'dromes were built, adding to the pre-war airfields at Three-Mile and Seven-Mile. Both Seven-Mile and Five-Mile (Wards) were the most heavily used by transport squadrons. At various times transport units were based at both fields, however as the war progressed Five-Mile became the headquarters for DAT and local C-47 units.

Introduction

Generals Dwight Eisenhower and Douglas MacArthur both credited unglamorous resources such as the bulldozer and the Douglas C-47 Skytrain as quintessential platforms to winning the war. This same claim was arguably even more relevant for the C-47 in the Southwest and South Pacific theatres, especially in mountainous New Guinea where the terrain was ready-made for air transport. Indeed, aside from their incessant transport duties, C-47s were directly involved in two major combat operations in New Guinea: the defence of Wau and the paratrooper drop at Nadzab (see Chapter 1).

This volume focuses on the markings and histories of the twenty-one USAAF squadrons which operated by far the largest fleet of C-47s in the Southwest and South Pacific. From humble beginnings in the first months of 1942 when a handful of C-53s had arrived in Australia by ship, the fleet grew rapidly from late 1942 as factory fresh C-47s were ferried across the Pacific by newly raised squadrons. During 1943 this transport fleet saw much hard use with many accidents occurring. Then throughout 1944 most of these squadrons transferred north to the Philippines, although many aircraft remained frequent visitors to Australia right up until the end of the war in August 1945.

USMC units operated the C-47 throughout the South Pacific, principally in support of the Solomons campaign, with their aircraft designated as R4Ds. Other regional users were the RAAF, RNZAF, RAF and Netherlands East Indies transport units.

While the commonly understood stereotype might be of homogenous and sterile Olive Drab C-47 airframes, closer inspection reveals that markings details were both complex and ever-changing. Many of the USAAF C-47s were named by their crews and received highly individualised associated artwork.

I hope readers enjoy the 100 profiles and associated information in this volume, which gives colour and context to the many C-47s which criss-crossed southern skies so many years ago.

Michael John Claringbould
Canberra
February 2022

Transport Squadrons
Command Structure SOPAC & SWPA
1942 - 45

FIFTH AIR FORCE

317th TCG	374th TCG	375th TCG	433rd TCG
39th TCS	6th TCS	55th TCS	67th TCS
40th TCS	21st TCS	56th TCS	68th TCS
41st TCS	22nd TCS	57th TCS	69th TCS
46th TCS	33rd TCS	58th TCS	70th TCS

(seven 33rd TCS C-47s served USN Oct/ Nov 1942)

(57th TCS seconded to 13th AF for month of April 1944)

THIRTEENTH AIR FORCE

US Marines	US Navy	RNZAF	403rd TCG
VMJ-152	MAG-25	No. 40 Squadron	13th TCS
VMJ-153	COMAIRPAC	No. 41 Squadron	63rd TCS
VMJ-253			64th TCS
			65th TCS
			66th TCS

(65th & 66th TCS transferred to 433rd TCG 9 Nov 1943)

The overall command structure of transport squadrons in the Fifth Air Force (SWPA) and Thirteen Air Force (SOPAC) during 1942-1945. Many other squadrons operated under Directorate of Air Transport control in the SWPA during this period.

CHAPTER 1
Overview of C-47 Operations in the Southwest and South Pacific

The conduct of transport operations throughout the Southwest and South Pacific theatres was both complex and costly. It was complex because the SWPA command system originated from a hybrid civilian/military structure with multiple commands and nationalities which was reformed into a larger version. This underwent numerous metamorphic changes, and as if this were not complex enough, C-47 markings were further complicated by consistent transfers of aircraft between units, and sometimes even air forces.

It was costly because C-47 operations incurred many losses, contradicting the post-war myth that transport losses in the Pacific were relatively light. Many fell foul to the harshness of the environment, and most C-47s suffered two or three major accidents during their time in the theatre. Combat losses by contrast, and against exaggerated claims at the time of alleged dangers involved, were comparatively light. In fact, there were few C-47s downed by enemy fighters. Numerous C-47s remain missing in New Guinea's mountains or were swallowed by the vast Solomons Sea. Some wrecks were located post-war however curiously their crew remain *in situ*: the very extreme geography which claimed them has deterred recovery. The comparative extent of USMC R4D losses, certainly in terms of human lives, is overwhelming.

Southwest Pacific Area (SWPA) and Department of Transport (DAT)

The origins of military transport in the South Pacific commenced in Australia with the creation of Air Transport Command (ATC) on 28 January 1942 at RAAF Amberley, to the west of Brisbane. With an initial modest complement of fourteen officers and nineteen enlisted men, in early February 1942 the ATC delivered materiel to bases in Java. Then within a short time it was assisting with evacuation of the same bases when the Japanese invaded.

On 3 April 1942 the ATC was redesignated as the 21st and 22nd Transport Squadrons, USAAF, with headquarters for each squadron based at Archerfield (Brisbane) and Essendon (Melbourne) airports. Both squadrons were bolstered with the addition of veteran pilots from the Philippine and Java campaigns, together with a reinforcement cadre from the US.

The ATC was replaced by the Directorate of Air Transport (DAT) which proceeded to become the fulcrum for all military transport operations in the SWPA, operating both USAAF and RAAF aircraft, with a sprinkling of Dutch crews. The command was renamed DAT in order to avoid confusion with the US Air Transport Command then servicing trans-Pacific routes, and with which it bore no relation. Group Captain Harold Gatty was later appointed its first commander, Gatty being a civilian specially commissioned into the RAAF in order to legitimise his military authority. DAT initially operated an eclectic suite of aircraft, including C-39s, C-49s, C-50s, C-53s, DC-5s, C-56s and C-60s, many of them ex-Dutch civilian and military aircraft which had evacuated from Java.

It should be noted that until the formation of the Fifth Air Force in September 1942, all SWPA air operations fell under the American-Australian Allied Air Forces command. While the RAAF had no air transport squadrons of its own at the start of the Pacific War, there was a sizeable civilian fleet of airliners which included DC-3s. These were often mobilised for military use, especially during the first year of the war. For this reason, and also because the main aerodromes used by the transports in Australia were civilian, military transports were issued civilian "VH" prefix radio callsigns which matched the "VH" prefix registration numbers used by Australian civilian aircraft. These callsigns, issued by DAT, appeared on all RAAF (and Dutch) transports which operated in Australia during the war, and on some USAAF aircraft, particularly in 1942 (these DAT callsigns are further discussed in Chapter 23).

From October 1942 the first of two additional USAAF transport squadrons arrived in Australia equipped with new C-47s. Within the following months the SWPA military transport fleet would overwhelmingly operate the C-47 as new units arrived and the original heterogenous fleet dwindled. The RAAF received its first C-47s in March 1943, and the new type represented a quantum increase in capability over the likes of DH-84 Dragon biplane transports.

Meanwhile, DAT underwent various changes following the formation of the Fifth Air Force, and various units fell in and out of its control. In due course DAT was primarily responsible for transporting freight and personnel within and Australia and to Port Moresby. Further north various USAAF troop carrier squadrons, equipped with C-47s, would operate under direct Fifth Air Force control in New Guinea, as explained below.

The first of several USAAF Troop Carrier Groups to serve in the SWPA was formed in November 1942 when the 374th TCG absorbed the original transport squadrons in the theatre. These were the 21st and 22nd Troop Carrier Squadrons which had been operating since April 1942 as noted above, together with the 6th and 33rd TCS which had arrived in Australia in October.

The 433rd TCG was the last transport group to arrive in the theatre, in late August 1943, with the 67th, 68th, 69th and 70th Troop Carrier Squadrons. However, both the 65th and 66th TCS were transferred from the 403rd TCG (assigned to SOPAC) to the 433rd TCG on 12 November 1943, making the 433rd the largest TCG in the USAAF, equipped with six squadrons. Its executive officer, Lieutenant Colonel Joseph Anderson, had previously been a captain with United Airlines for ten years.

In order to coordinate transport movements in combat areas between the increasing numbers of units appearing in the theatre, the 54th Troop Carrier Wing (TCW) of the Fifth Air Force was formed on 13 May 1943 as an additional command structure to DAT. This meant troop carrier groups sometimes rotated between these two command structures. An example of this is when on 1 October 1943 the long serving 374th TCG was given respite from combat and transferred from the 54th TCW back to mainland Australia and returned to the control of DAT. Further elaboration on DAT, its role and markings can be found in Chapter 23.

Transport units of course had to deal with their own logistics, and the pattern for USAAF units transferred to the theatre from the US followed a similar pattern: the air unit would fly across the Pacific route whilst its support contingent would follow by sea. This involved a lot of "catch-up" entailing a few months of self-support by the air unit. A good example is the 433rd TCG whose ground echelon sailed from San Francisco on 25 September 1943 by which time their C-47s had already been flying in the theatre for a month. They eventually arrived aboard the USS *West Point* at Sydney on 10 October before heading north by train to Brisbane and then Townsville. It then took 74 planeloads to fly them to Port Moresby from Townsville. In the interim the group's air detachment of four squadrons had relied on other units for logistics support.

Miscellaneous Units

Note that later in the war the 2nd Combat Cargo Group comprising the 5th, 6th, 7th and 8th Combat Cargo Squadrons also served under the 54th Troop Carrier Wing. After inauguration and training at Baer Field, Indiana, these units completed a move to Biak by the end of November 1944. They then operated mainly in the Philippines but some of their limited C-47 inventory also appeared in Australia and even ventured into the Admiralties (north of New Guinea) in this timeframe. The group, as part of the Far East Air Force, moved to Leyte in May 1945 and then Okinawa in August 1945. Their C-47s are not represented in this volume for two reasons. First, these squadrons mainly operated the C-46, and second, their appearance in the SWPA was the exception and not the rule.

The 3rd Air Cargo Group was formed later in the war and mainly saw service in the Philippines. Unusually, the 318th TCS was administratively attached to this group as an orphan outfit, and its C-47s sometimes appeared at Nadzab in late 1944 and 1945. However, this squadron is not included in this volume either due its only occasional SWPA activity.

Air Transport Command aircraft often appeared in the theatre, however due to the theatres' distances they frequently materialised in the shape of four-engine C-54s. ATC aircraft were forbidden from entering "combat zones" incurring the alter-acronym "Allergic To Combat" from their regular air force counterparts. ATC's Pacific Wing was established as the USAAF 25th Ferry Wing on 27 June 1942, before being redesignated the South Pacific Wing on 1 July 1942 and then becoming the Pacific Wing in January 1943 when it moved its headquarters to Hawaii. Its aircraft flew the South Pacific air route to Brisbane or Williamstown, Australia, from Hamilton Field, California, via Hawaii, Nadi (Fiji) and Noumea transporting cargo and passengers. Later in the war it also flew to New Zealand via Henderson Field once this had been classified as a "non-combat" area. None of its C-47s are represented in this volume as they were not based in either the Southwest or South Pacific theatres.

Finally, the Combat Replacement Training Centre, also known as the 360th Service Group, operated at Nadzab in late 1944/early 1945 commanded by the former 3rd BG commander, Colonel "Jock" Henebry, to train incoming crews for combat. It deployed several C-47s transferred from a variety of other outfits for this purpose, some of which had "CRTC" stencilled on the nose in white. However, none of these aircraft are profiled in this volume.

South Pacific Area (SOPAC)

Transport logistics in SOPAC were initially conducted by Marine Aircraft Group 25 (MAG-25) to support the Guadalcanal campaign. This was reformed in late November 1942 as the South Pacific Combat Air Transport Command, giving birth to the widely recognised acronym SCAT. Organised as an autonomous command structure at the end of the Guadalcanal campaign at the direction of Vice Admiral Aubrey Fitch, SCAT was a conglomerate of both USMC and USAAF units. It grew from VMJ-253 and small MAG-25 detachments, together with the 13th TCS. By February 1943 SCAT had also acquired VMJ-152, SMS-25 and the USAAF 801st Medical Air Evacuation Squadron. Other additional units that also joined were VMJ-153, the 63rd TCS and the 64th TCS.

SCAT provided long-distance transport throughout the South Pacific theatre for personnel, munitions, rations, spare parts, medical supplies, liaison and VIPs. Returning flights carried wounded personnel, those travelling on leave to Australia or New Zealand, as well as administrative documents and components for overhaul, including engines. Medical flights typically carried a nurse, corpsman or flight surgeon as part of the crew.

C-47 Combat Operations in the SWPA

Aside from the incessant movement of men and materiel throughout the Pacific, C-47s were directly involved in two major combat operations in the SWPA: the defence of Wau and the paratrooper drop at Nadzab. On the other hand, there were no comparable major C-47 combat operations in the SOPAC theatre.

In January 1943 a plan to reinforce Australian troops at the mountain township of Wau by airlift occurred in direct response to a Japanese advance on the area. This was led by Major General Okabe Toru, commanding the IJA 102nd Infantry Regiment, supported by a field artillery battalion. The Japanese viewed Wau airfield as an ongoing threat to their nearby bases at Lae and Salamaua. Okabe was tasked with its capture primarily so it could be used as a staging base for an advance on Port Moresby. Allied intelligence confirmed the arrival of the Okabe Detachment at Lae on 7 January tasked to march on Wau. The following day the Australian 17th Brigade was ordered to Wau where it would assimilate with the Australian Kanga Force, already *in situ*, to meet the threat.

The five-day period from 28 January to 1 February 1943 marks the critical phase of the resultant Battle of Wau. USAAF C-47s set a record when they made 71 deliveries to Wau from Port Moresby on 31 January. With one exception, every one of the 35 transports involved made two return trips, delivering 355,000 pounds of men and materiel for the ongoing battle.

This transport effort tipped the balance in favour of the Allies. When ordered to withdraw from Wau, Colonel Maruoka Yasuhei reported that the Australians still controlled high ground south of the airfield. He also stated that in the past two days 130 transports had delivered reinforcements and supplies.

Following the landmark Battle of Wau, the 317th TCG returned to Australia to join its ground echelon which had since arrived in Australia aboard the SS *Maui* and set up at Garbutt Field outside Townsville. Plans were soon afoot for another major operation. At Mareeba the group commenced air-drop exercises with the 503rd Parachute Infantry Regiment around the Cairns region. An air-drop to seize Nadzab, leading to the capture of Lae, was set for September 1943. Although the 375th TCG would co-ordinate the drop, it was determined that the 317th TCG possessed the best qualifications to lead the operation. It spent its last three months in the US training with both parachute battalions and gliders. At Fort Benning the group had trialled numerous drop procedures from which it had developed, *inter alia*, a method of airspeed control which was adopted for the Nadzab operation. The C-47 crews and paratroopers developed a mutual trust during the Australian training missions as they refined coordination tactics. Airborne training around Cairns reached a peak throughout April to July 1943.

Sunday 5 September 1943 marked "Z Day" initiated at dawn by 27 crews from the 317th TCG who led the first airborne assault operation in the Pacific. At 0630 their C-47s warmed engines at Seven-Mile, Port Moresby, followed by formations from the 375th TCG and then from all of the other units. General Douglas MacArthur and entourage would witness the drop from a patrolling B-17. Colonel Paul Prentiss, the commander of the 54th TCW, flew as co-pilot in the lead C-47 *Honeymoon Express*. At 48 years of age, Prentiss was a career army officer from San Antonio, Texas, and had flown since the last year of the Great War in 1918. Fittingly, Major Bill Williams, the 317th TCG's operations officer, was the pilot in command. Photographers were everywhere.

Weather delayed take-off as cloud blocked the mountain passes of the imposing Owen Stanley mountains. Reports from the tasked B-25 "weather ship" were intermittent due to bad radio reception. Many of the aircrew carried good-luck keepsakes. Lieutenant Courtney Faught, a 25-year-old pilot, taped a photo of his infant son on the instrument panel. Faught was flying C-47 number twelve in the overall formation, named *The Broadway Limited*.

The drop was divided into three flights:

• A Flight was comprised entirely of C-47s from the 65th and 66th TCS led by Lieutenant Colonel John Lackey, the 54th TCW deputy commander, with co-pilot Major Don Smith, the commander of the 66th TCS.

• B Flight was led by Lieutenant Colonel Joel Pitts, the 375th TCG commander. It comprised all four squadrons of the 375th TCG (the 55th, 56th, 57th and 58th TCS).

• C Flight was led by Colonel Prentiss, as described above, and comprised the 41st and 46th TCS.

All 79 transports were airborne within twenty minutes, reflecting a tightly coordinated mass departure with take-offs every fifteen seconds. The gathering first flew southeast for 30 miles to allow all transports to get airborne, then turned back to complete the re-join over the SS *Macdhui*, a sunken ship in Port Moresby's Fairfax Harbour. The formation then banked northwest towards the main rendezvous over Rogers airstrip, some 30 miles away. Overhead Rogers the transports arranged themselves into a line of three-aircraft Vs at 9,000 feet. Numerous fighters flew ahead and above for protection. Around 0930 the formation flew through the Sunshine Pass into the Watut Valley towards Tsili Tsili where they commenced descent to 3,500 feet. Here each flight reformed into six-ship elements in right echelon formation, with all three columns abreast. This manoeuvre maximised use of the drop zone's wide valley, reducing the length of the formation and enabled closer fighter coverage.

At this juncture more fighters joined from Dobodura and Tsili Tsili airfields, with the overall armada now made up of 302 aircraft. The transports droned into the Markham Valley where they descended to a few hundred feet. Heat generated from the ubiquitous kunai grass generated turbulent thermals, buffeting the C-47s and their occupants. B-25s and A-20s broke off and strafed the dirt road along the valley floor which led to Lae. Other A-20As flew on both sides of the formation and laid down a thick white smoke screen. This rose a good thousand feet to hide the drop from enemy eyes.

At a marked bend in the Markham River perpendicular to the approach, the transports maintained 400 feet and slowed to 95 knots, conscious that any individual airspeed variation would spill into a domino effect. The sudden departure of the paratroopers as they jumped *en masse* substantially shifted the centre of gravity and the pilots had to wrest their control columns to maintain level flight. The drop lasted exactly four and a half minutes as planned. More than 1,700 parachutes descended into the kunai grass below. After clearing the three-mile-long drop zone, the transports then descended to 50 feet where crews retracted static lines, including those which had released equipment bundles. After all lines had been secured, the all-clear was broadcast and the planes accelerated and climbed. A and B Flights returned directly to Port Moresby around midday, however C Flight first proceeded ten miles to the east where they dropped decoy dummy parachutists from 1,000 feet.

Thus concluded the first successful Allied airborne assault of the war. Casualties were limited to two parachutists whose equipment failed. Another was safely stuck in a tree but fell to his death when he released his harness, and 33 others sustained a range of injuries. The C-47s encountered no enemy fighters or anti-aircraft fire. When four transports reported tracers blocking their flightpath it turned out to be from strafer B-25s. With the valley secure, the first of five eventual airfields was quickly prepared at Nadzab. Within the next six days continuous C-47 flights had delivered 420 planeloads of men and materiel to the new strip.

The early batches of USAAF contract C-47s had this unique stencil style applied at the Long Beach production plant as illustrated here. Production lines at Santa Monica, California, and Oklahoma City, Oklahoma, also produced a variety of different stencil styles.

CHAPTER 2
C-47 Markings in the SOPAC and SWPA

This volume concentrates on Douglas transports, mainly the C-47 and R4D series, which were the overwhelmingly numerous transport type to serve in the Southwest and South Pacific theatres. The scope also includes a small number of ex-civilian DC-3s which found their way into USAAF service, redesignated as C-50s, plus a small number of C-53s.

No publication to date has adequately defined the complex and ever-changing markings systems which applied to these transports. Some standardised systems did exist; however, they can be challenging to define due to ongoing exchanges of inventories, different command structures, and frequent reassignments of airframes through service squadrons or loans to other units.

This publication focuses on aircraft markings only in the Southwest and South Pacific theatres. These further evolved to varying degrees when USAAF transport units left these theatres for the Philippines, a different and distant war compared to the one fought from 1942 until early 1944. Several examples of these later Philippine-era markings are included in this volume as these aircraft often returned to Australia and the South Pacific right up until the cessation of hostilities.

The biggest inventory swap occurred when the 317th TCG arrived in Australia on 15 January 1943 bringing 52 new C-47s and fresh crews. The group was initially attached to the 374th TCG and advanced to Port Moresby to support operations at Buna and Wau. The 374th integrated 317th TCG crewmembers shortly thereafter, uniting their units to full strength at a critical time when the battle for Wau demanded maximum resources. After the pressure of the Wau campaign had subsided, the entire 317th TCG fleet of C-47s was transferred to the 374th in exchange for their war-weary fleet which the 317th TCG took back to Australia.

Therein lies one of several key reasons behind many markings misunderstandings in the theatre. The 317th TCG C-47s arrived in Australia marked with three-digit stateside squadron numbers in the 400s, 500s, 600s and 700s series. These were relatively small in size and stencilled to the nose of the aircraft. Furthermore, after these airframes were allocated to the 374th TCG, many of these numbers remained extant on the airframes throughout 1943. In many cases they remained on the airframe after the application of new squadron numbers.

Markings - General

Throughout both the SOPAC and SWPA theatres nearly all C-47s and R4Ds were painted overall Olive Drab with a neutral grey under surface. Nearly all had random factory applied Forest Green patterns sprayed on the leading edges of the wings, fin and tailplane. These random patterns, applied with a spray gun over rubber stencils, sometimes over-sprayed onto wing under surfaces. Olive Drab faded to varying degrees depending on its level of exposure. Control surfaces appeared lighter in colour even when the C-47 was brand new, due to the lower reflection of canvas and also because ancillary control surfaces were built, doped and then painted by subcontractors. Early production C-53s were painted overall Forest Green instead of Olive Drab. Dutch and Australian C-47s came up with all sorts of greens, depending on what was in stock in the relevant workshop.

Faded Olive Drab became browner due to harsh tropical wear, not reddish or pink (as one author has creatively claimed, presumably misinterpreted from faded Kodachrome slides). Heavy wear and grease stains permeated door, cowl and control hinges. Modellers can take satisfactory refuge in the fact that repairs using fresh paint and the widespread practice of cannibalised parts meant that, in practice, every C-47 and R4D was different. Exhaust stains were also ubiquitous and became a permanent feature long after engine refits. Fuselage interiors were sprayed with etch primer which was an outstanding zinc chromate green. C-47s on dedicated missions often had the jump door removed and/or a temporary number chalked or painted in water colour on the fuselage side to assist with loading.

Both the USN and USMC officially disapproved of personalising aircraft, however there are many cases where artistic creations briefly appeared for photo opportunities, only to disappear when the shutter had done its work. R4Ds in both theatres, converted from USAAF contracts, appeared in Olive Drab schemes and as such often continue to be captioned and identified as C-47s in all manner of publications. This has had the unintended post-war consequence of detracting from the USMC's massive contribution to SOPAC transport operations and thus unfortunately diluting their historical significance. In observance of the discretion of USN and USMC markings practice, R4Ds retained only a BuAer number in small stencils on the fin, usually in black but sometimes applied in white.

Natural metal finish C-47s started appearing late in the war at first in the shape of new USMC R4Ds and RAAF C-47s, however the majority of 54th TCW C-47s retained their Olive Drab camouflage into the post-war era. Several FEAF C-47s with command responsibilities appeared in natural metal finish, and such schemes incorporated a black anti-glare panel just forward of the windscreen.

Nose art usually appeared on the port side in both theatres, however sometimes it appeared on both sides, and even with different names. Some squadrons adopted nose art themes, the most telling of which was the 41st TCS which named many of its transports after trains in the US. Some squadrons had customised logo adhesive decals made in Sydney, which were later applied to the aircraft on one or both sides of the forward fuselage.

Once the Philippine campaign commenced, two key changes were made. First, a prefix of X or W was added to existing squadron numbers. This was not done to differentiate between the newer C-46 Commandos coming online, as often claimed (the C-46s bore identical prefixes) but was added to the callsign to differentiate between transports and other types of aircraft. Secondly, several groups started using airline-style fuselage headings such as *Jungle Skippers* (317th TCG), *The Tokyo Trolley* (375th TCG) and *Frontline Airline* (433rd TCG). Nonetheless the use of such airline-style titles was adopted only later in the war, after the units had transferred to the Philippines, and was thus not a feature of SOPAC or SWPA operations.

SQUADRON NUMBERS
US Marine & USAAF Transport Squadrons
South & SW Pacific 1942-45

DIRECTORATE OF TRANSPORT
Allocated radio callsigns VHCAA to VHCZZ
(C-47 series VHCAT to VHCXM)

US Marine VMJ Squadrons
Last two digits of Buer
applied to tail, fuselage or engine cowl
(e.g Buer 12323 becomes 23, Buer 4706 becomes 06)

317th TCG
39th TCS (1-25)
40th TCS (26-50)
41st TCS (51-75)
46th TCS (76-99)

(added prefix **'X'** in late 1944)

374th TCG
6th TCS (51-75)
21st TCS (1-25)
22nd TCS (26-50)
33rd TCS (76-99)

(added prefix **'W5'** late 1944)

375th TCG
55th TCS (101-125)
56th TCS(126-150)
57th TCS (151-175)
58th TCS (176-199)

(added prefix **'X'** in late 1944)

433rd TCG
67th TCS (301-325)
68th TCS (326-350)
69th TCS (351-375)
70th TCS (376-399)

(added prefix **'X'** in late 1944)

403rd TCG
13th TCS (1-35)
63rd TCS (36-70)
64th TCS (71-85)
65th TCS (201-225)
66th TCS (226-250)

(seconded to
433rd TCG
9 November 1943)

A table of squadron number ranges used by USAAF transport squadrons. It also includes the DAT callsign range and an explanation of USMC squadron numbers.

In summary, the markings system for transports operating in the SOPAC and SWPA theatres can be divided into the following themes:

• USAAF squadron numbers were normally painted or stencilled in yellow or white immediately behind the cockpit window. Both the Fifth and Thirteenth Air Forces tried to standardise this system, however exceptions exist, usually when previous numbers were retained following airframe exchanges.

- 317th TCG three-digit squadron numbers on the nose which had been applied in the US, but which remained on newly arrived airframes as explained above.

- Directorate of Transport "VH" radio callsigns, applied to the fin or fuselage or both; USAAF serial numbers were sometimes painted over in the process.

- USMC R4D units applied the last two digits of the BuAer number to the fin, fuselage or engine nacelle, or a mixture thereof.

- RAAF units applied their own squadron codes in the form of large letters on the fuselage.

- Several USAAF squadrons applied squadron insignia behind the cockpit, either on one or both sides.

- Around mid-1944 some USAAF squadrons colour-coded their rudder trim tabs.

The end result is that many transports wound up conducting operations with a mixture of current and legacy markings. These were applied with the ideal of consistency, but often wound up with overlays of multiple units.

The Douglas C-47 production line at Long Beach, California, in 1942. Serial numbers are stencilled on the tail fin. The foremost aircraft is serial number 41-18625 which wound up serving with the 63rd TCS in the Solomons.

These C-47s are fresh off the Santa Monica production line in early 1942. The foremost is serial 41-18415, which wound up in New Guinea by year's end serving with the 46th TCS. After surviving the war and an extensive post-war civilian career, it was finally lost to a crash on 30 May 1978 at the coastal airfield of Santo Tomas in Guatemala.

Quotations

I flew several times across the range in these transports. But I can't remember any trip when my stomach didn't feel as if it were doing slow rolls or when the hair at the nape of my neck was not bristling with fear. I could never get accustomed to driving through a grey rain cloud, seeing a vaguely darker shape ahead, realising sickeningly that it was the side of a mountain wall just as the plane lurched violently and nearly rolled over as it turned to get out of trouble ... The men who flew the transports crossed the range six, eight or ten times a day ... that took guts and stamina and morale and willpower and all the other things that are easy to write about. Yet the main topic of conversation among these kids was how much stuff they could get through to the troops.

Australian war correspondent, George Johnston

You think differently in the combat zone, and that means acting and being different. Our new, clean airplanes without mission symbols painted on them, or identity numbers looked naked. And like boats, bow-wows and books, airplanes need names. They're a cold piece of machinery, but to those of us who have used them, loved them and trusted our lives to the amalgam of metal and synthetics, they have a personality and a gender. And how could that gender be anything other than female? You talk to them, cherish them, stroke them, and take better care of them than you do of yourself. In our predisposition to humanize our machines, we gave them names, often accompanied by an illustration of a pinup girl, a Vargas or Petty girl with impossible anatomical proportions, but reflecting the sentiment or, more likely, the fantasy of a crew member.

68th Troop Carrier Squadron C-47 pilot Lieutenant Robert Stenglein

The subject of Profile 2, C-47 Dear Mom, at Ward's 'drome with its crew.

C-47 serial 41-18697 Norma of the 6th TCS is unloaded at Bulolo, likely during the Wau campaign of January 1943. As noted in Profile 4, this ex-317th TCG aircraft was soon renamed Yard Bird.

CHAPTER 3
6th Troop Carrier Squadron "Bully Beef Express"

The 6th TCS was led by Captain Hamish McClelland when it departed the US mainland on 2 October 1942 with thirteen C-47s. It thus became the first USAAF transport squadron to cross the Pacific, its C-47s arriving at Port Moresby via Hawaii on 13 October 1942. They were followed by the ground echelon which arrived at Townsville some five weeks later. The 6th TCS was assigned to the 374th TCG along with another three squadrons which had also arrived or been formed in the SWPA by this time: the 21st, 22nd and 33rd TCS.

McClelland was replaced as 6th TCS commander on 22 May 1943 by Major John Lackey, who was replaced in turn on 3 December 1943 by Captain William Peterson.

During its time in the SWPA the 6th TCS lost eight C-47s to operational causes and one to enemy aircraft. Its last loss in the theatre occurred on 1 November 1944 when one of its C-47s caught fire while being refueled at Dobodura.

Markings

In New Guinea the 6th TCS called itself the *Bully Beef Express* and adopted the logo seen on page 24, which continued to be used after the war. It was allocated squadron numbers 51 to 75, which were applied from mid-February 1943 onwards after an inventory swap with the 317th TCG. Some of these squadron numbers appeared concurrently with the three-digit ones applied to all 317th TCG C-47s in the US, although several were painted over when the new squadron numbers were issued. In addition to the C-47s profiled here, the following names also appeared on 6th TCS C-47s: *Boobs, The Broadway Limited, Cheryl, Irish, Johnny Reb, Johnny Reb II, Kentucky I Love Thee, Miss Alabama* and *Norma*.

The subject of Profile 3, C-47 Irene, at Mareeba, Queensland.

6th TCS

Profile 1 - C-47 serial #41-38601, *Swamp Rat*, (no squadron number)

This aircraft is profiled as it appeared on 7 November 1942 when it became the first transport to land on a freshly cleared strip at Pongani on Papua's northern coast. The reason for the horizontal white fin stripe is unclear, it was perhaps a formation marking. The aircraft was lost before it had a squadron number allocated. An official photographer took several photos of this transport as provisions and ammunition were being unloaded for troops of the US 32nd Infantry Division, fighting at nearby Buna. The transport was temporarily bogged before finally returning to Wards 'drome at Port Moresby. For the next three weeks it continued to ply goods between Wanigela, Dobodura, Pongani and Port Moresby.

Just before 0900 on 26 November 1942 it departed Dobodura for the return trip to Wards. Four crew were onboard including the pilot Lieutenant Earl Lattier. Observing the doctrine of safety in numbers, Lattier joined forces with another C-47, *Shady Lady*, crewed by pilot Staff Sergeant Marvin Brandt and three others from the 33rd TCS. While climbing over foothills, both transports were intercepted by six No. 252 *Ku* A6M2 Zeros and shot down about five miles south of Dobodura. The culprits were led by Flying Petty Officer First Class Kobayashi Katsutauru who claimed two "Daugasu" transports. The six Zeros fired a bare minimum of 360 x 7.7mm and 60 x 20mm rounds to down both C-47s, reflecting skilled gunnery.

Profile 2- C-47 serial #41-18673, *Dear Mom*, squadron number 56

Transferred from the 317th TCG in late February 1943, *Dear Mom* was assigned to pilot Lieutenant George Beaver's crew, including co-pilot Hubert Bronson, engineer Staff Sergeant Kenneth Wagoner and radio operator Corporal Lawrence Billmaier. It is profiled as it appeared at Wards 'drome in April 1943.

Profile 3 - C-47 serial #41-18646, *Irene*, B Flight

Assigned into the 6th TCS in January 1943, *Irene* is profiled as it appeared at Mareeba when it served with B Flight at the end of 1943. Unusually, the original squadron number 56 was painted over and replaced by the letter B for this purpose, while the original name *Irene* was repainted in different calligraphy. On 3 November 1942 the transport made its most eventful flight when it made the last of five supply-drop runs over Kokoda strip. The ropes retaining the main cargo door open broke during the drop, tearing it off its hinges and catching on the tail plane. When the aircraft landed back at Port Moresby the door fell away as it touched down.

Profile 4 - C-47 serial #41-18697, *Yard Bird*, squadron number 57

This transport was previously assigned to the 317th TCG with which it arrived in theatre with squadron number 704 and the name *Norma*. *Yard Bird* is profiled as it appeared in late March 1943 at Wards 'drome with these previous markings painted over.

Wounded being loaded at Henderson Field aboard Lady Eve, the subject of Profile 7. The white stripe on the tail is prominent and is possibly a formation marking.

CHAPTER 4
13ᵗʰ Troop Carrier Squadron "The Thirsty 13ᵗʰ"

The 13ᵗʰ Troop Carrier Squadron arrived at Plaine des Gaiacs airfield on New Caledonia on 9 October 1942, about halfway through the Guadalcanal campaign. It was the first of five transport squadrons assigned to the 403ʳᵈ TCG. Nicknamed *The Thirsty 13ᵗʰ* and without having to negotiate dangerous mountainous terrain like their New Guinea counterparts, the squadron attained a remarkable safety record. During its time in the South Pacific it had three headquarters, firstly when it arrived at Plaine des Gaiacs; then Tontouta (also on New Caledonia) from 19 December 1942; and finally Espiritu Santo in the New Hebrides from 23 October 1943. The 13ᵗʰ TCS moved its headquarters to Biak in October 1944, marking exactly two years in the South Pacific. It was the first USAAF squadron to join SCAT.

By mid-October 1942 an acute aviation fuel shortage had developed on Guadalcanal which threatened to curtail fighter operations, and the 13ᵗʰ TCS ferried urgent supplies to Henderson Field. That month alone, SCAT units including the US marine R4Ds delivered around 105 tons of ammunition, petrol and supplies. They also hauled five tons of mail and 339 passengers, as well as 498 wounded who were evacuated to Espiritu Santo. During this time SCAT transports were sometimes departing with take-off weights in excess of 30,500 pounds, far in excess of the 25,200 pound limitation as endorsed by the US Civil Aviation Authority. Acting as navigation escort aircraft, SCAT pilots also took on a navigation role, guiding 65 combat planes to Guadalcanal in October 1942 alone.

Captain Erling Nasset was appointed as the new squadron commander in July 1943 after flying with the 13ᵗʰ TCS for eighteen months. The squadron only lost two aircraft during its entire South Pacific service – one was hit by shellfire over Guadalcanal causing it to later ditch, and another fell to operational causes. The squadron's pilots flew to the most diverse number of destinations in the South Pacific, from New Zealand to New Guinea; *inter alia* Plaine des Gaiacs, Tontouta, Magenta, Henderson Field, Auckland, Nadi, Segi Point, Brisbane, Sydney, Espiritu Santo, Efate, Ondonga, Munda, Torokina, Emirau, Saidor, Treasury Island, Kiriwina, Nadzab and Los Negros.

In addition to the C-47s profiled here, the following names also appeared on 13ᵗʰ TCS C-47s: *Baby Shoes, Billie, Bastard, Our Eleanor, Col Bud II, Commanche Bell, Chuggar, Connie, Daisy, Flyin' Jenny, Green Goggled Ghost, I'm a Comin', Karen, Kee-bird, The Lana T, L'il Joy, Mickey McGuire, Old Boulder, Peck's Beau, Pluto, Pudgie, Ramblin' Wreck, Risky, Risky II, The Rod, Sally, Snafu, The Tar Heels, Teenie* and *The Wolf*.

13th TCS

5

6

7

8

Profile 5 – C-47 serial #41-18654, *Cat Fish*

This aircraft was included in the original batch assigned to the 13th TCS and had nose art applied to both sides of the nose. It is the only C-47 in the South Pacific theatre to have shark-teeth markings applied.

Profile 6 – C-47 serial #41-18572, *The Sad Sack*

This aircraft was another among the original batch assigned to the 13th TCS. It had similar nose art on both sides of the nose.

Profile 7 – C-47 serial #41-18580, *Lady Eve*

This transport was named in honour of Eve, the girlfriend of its first assigned pilot Lieutenant Don Bergstrom. He further developed the name into *Lady Eve* after the 1941 film titled *The Lady Eve*. This was a film comedy starring Barbara Stanwyck and Henry Fonda, based on a story about a mismatched couple who meet during an ocean cruise.

Profile 8 – C-47 serial #42-23605, *Sweet Leilani*

This transport was named by it crew chief Sergeant Joseph Lalonde in early September 1943, shortly after it was assigned to the 13th TCS. He wrote into his diary on 3 September 1943:

> … I have been working on my plane quite a lot lately, fixing it up, and putting a name on it. Here it is: *Sweet Leilani*. It looks better on the ship. It's quite famous now, as it has carried some famous people. It also set a new squadron record for hours flown in one month. I am quite proud of it.

13th TCS

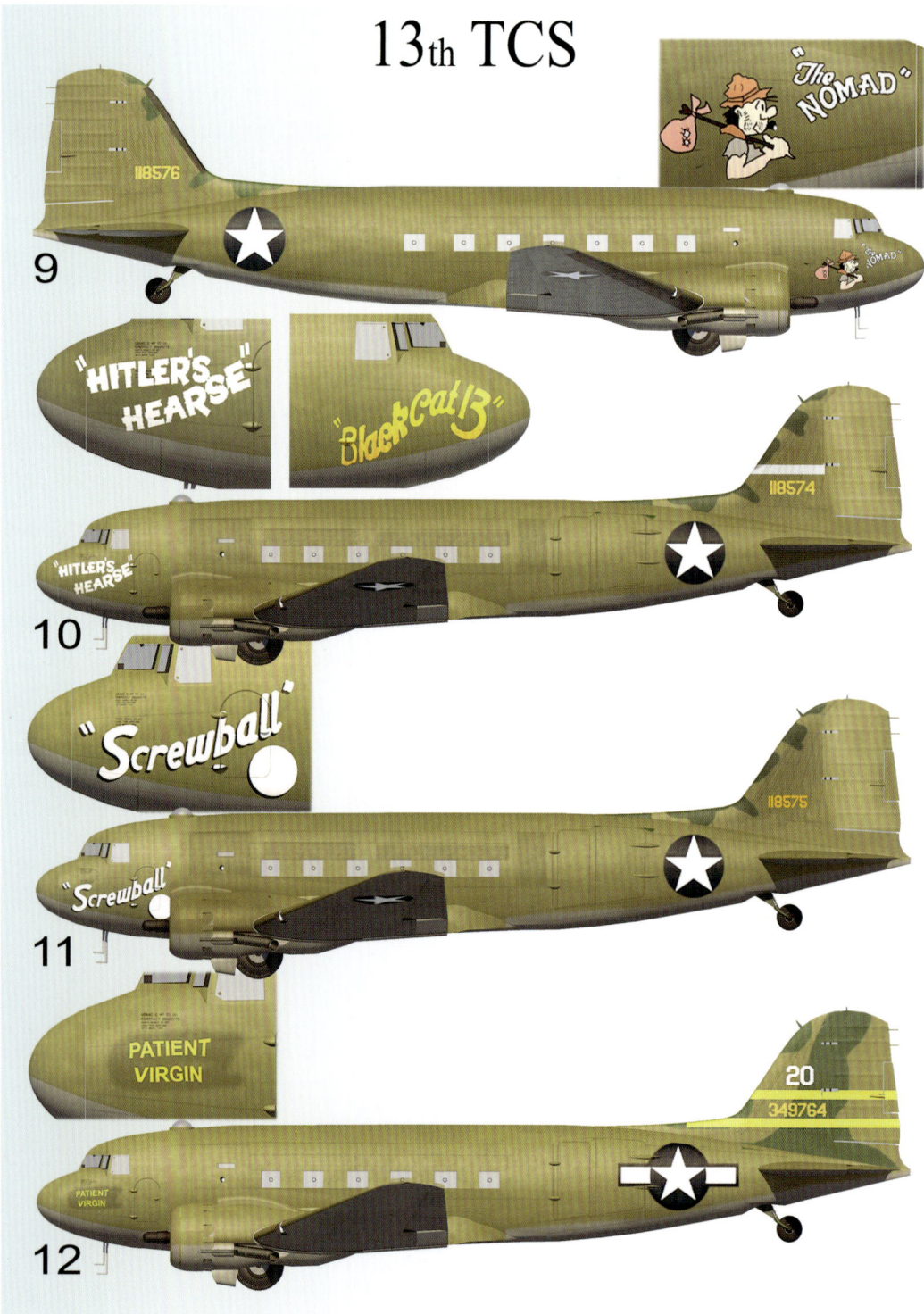

"The NOMAD"

9

118576

"HITLER'S HEARSE"

"Black Cat 13"

"HITLER'S HEARSE"

118574

10

"Screwball"

"Screwball"

118575

11

PATIENT VIRGIN

20
349764

PATIENT VIRGIN

12

Profile 9 – C-47 serial #41-18576, *The Nomad*

This aircraft was another among the original batch assigned to the 13th TCS on 5 October 1942. It only carried art on the starboard side.

Profile 10 – C-47 serial #41-18574, *Hitler's Hearse / Black Cat 13*

This aircraft was among the original batch assigned to the 13th TCS on 5 October 1942. It carried the name *Hitler's Hearse* on the port side and *Black Cat 13* on the starboard side.

Profile 11 – C-47 serial #41-18575, *Screwball*

This aircraft was among the original batch assigned to the 13th TCS on 5 October 1942, carrying art only on the port side.

Profile 12 – C-47 serial #43-49764, squadron number 20, *Patient Virgin*

Assigned into the 13th TCS in late 1944, the markings on this transport reflect the Philippine era, however the aircraft sometimes transited the Australia/South Pacific region in late 1944/early 1945. By mid-1944 the 13th TCS commenced allocating squadron numbers, and the double yellow stripes were applied to all squadron aircraft as a unit marking.

C-47 Old Boulder (serial #42-93506) joined the 13th TCS late in June 1944. It was transferred to No. 34 Squadron, RAAF, some time in 1945, and is seen here shortly after the transfer with a painted kangaroo to the rear of the cockpit.

Pudgie on the 13th TCS flight line. Note the star on the nose.

C-47A Homing Pidgeon serial number 42-23417 was briefly assigned to the 13th TCS on 9 May 1943, before transfer to the 13th Air Depot where it is seen in New Caledonia.

The port side of Profile 6, The Sad Sack, on display over the Solomons.

Snafu, another 13th TCS original C-47, at Tontouta, New Caledonia.

The ground crew of C-47 serial #41-18567 Clay Pigeon, the subject of Profile 13, pose with their aircraft at Wards 'drome.

CHAPTER 5
21st and 22nd Troop Carrier Squadrons

The history of these two squadrons is closely entwined, and thus best summarised in the same chapter. On 3 April 1942 Air Transport Command in Australia was ended as an entity and in its place the 21st and 22nd Transport Squadrons (TS) were created, with headquarters for the former moved from Melbourne to Archerfield in Queensland a few weeks later. The strength of both squadrons was bolstered with an influx of veteran pilots who had flown in the Philippine and Java campaigns, together with a cadre of enlisted men from the US.

21st Troop Carrier Squadron

On 22 May 1942 the 21st TS made its first operational flight in New Guinea carrying troops into Wau and Bulolo. It continued to operate between Port Moresby and Wau at various intervals with Airacobra fighter cover. Then after Japanese troops landed at Buna in late July 1942, the squadron was one of a handful which helped deliver supplies and reinforcements for Australian troops fighting in the mountains on the Kokoda Track. From August some of these supplies were air-dropped to specially prepared drop zones.

On 26 July 1942 both the 21st and 22nd TS were redesignated as Troop Carrier Squadrons (TCS). The first large scale troop movements by these squadrons occurred in September when three regiments of the US 32nd Division were transported to New Guinea from Brisbane and Townsville. Commencing in mid-October a concentrated effort was made to deliver supplies and troops for the final Allied drive on Buna.

The 21st TCS operated a mixture of C-39s, C-49s and C-53s in addition to the ubiquitous C-47 and finished the war serving in the Philippines alongside the 22nd TCS. Thus, the pair became the two longest-serving squadrons in the wider South Pacific theatre. The 21st TCS lost seven transports in the SWPA to operational causes, and two to JAAF fighters. Major Edgar Hampton was the squadron's first commanding officer, retaining command until 12 October 1942, when he was replaced by Major Fred Adams.

On 12 November 1942, both the 21st and 22nd TCS were reassigned into the newly designated 374th TCG Group, and on 29 January 1943 both squadrons were based at Wards 'drome, Port Moresby. During its SWPA service the 21st TCS was allocated squadron numbers between 1 and 25. The squadron started painting its trim tabs white in early to mid-1944 as a squadron identifier. It came to call itself *The Beeliners*, a name which carried into post-war service.

In addition to the C-47s profiled here, the following names also appeared on 21st TCS C-47s: *The Abortion, Galahad, Airline Algie, Andrea, The Apple Cart, Battling Bishop, Calamity (Mary) Jane, Contrary Mary, Down & Go, Eager Elaine, Foitle Moitle, Geronimo, Lakanookie, Leone, Liliane, Maxine, The Pacemaker, Quitchyr Bichen/Flamin Mamie* and *Touch of Texas*.

The subject of Profile 14, C-47A-15 serial #42-92802 Fair Dinkum, unloading cargo at Goodenough Island.

A line-up of 22nd TCS transports on the grass field at Dobodura, New Guinea.

22nd Troop Carrier Squadron

When Air Transport Command was redesignated into the 21st and 22nd TS on 3 April 1942, the 22nd TS was temporarily commanded by Lieutenant Francis Feeney with its headquarters established at Essendon, pending arrival of a more senior officer. By the end of May 1942, it was operating an eclectic collection of former Dutch airliners and military cargo aircraft, including Lockheed Model 14s, DC-5s, DC-2s and DC-3s.

The 22nd TCS perhaps underwent more leadership change than any other transport squadron in the SWPA: Captain Raymond Swenson superseded Feeney on 2 May 1942 when he arrived from the US, followed by Major William Bradford on 21 May 1942. Then, on 22 July 1942, shortly after Feeney had been promoted to major, he again assumed command which he retained until 6 April 1943 when he was transferred to 374th TCG Headquarters and Captain Pierre Jacques took over. On 30 April 1943 Jacques was ordered back stateside and replaced by Captain Fred Henry. On 30 May 1943, he too was ordered back to the US, whereupon Captain Perry Penn assumed command. When Penn departed for the US on 26 September 1943, Major Robert Beebe from 374th TCG Headquarters took over and he led the unit throughout the remainder of its SWPA tour.

The 22nd TCS moved from Essendon to Garbutt near Townsville on 17 September 1942. Then on 12 November 1942 both the 21st and 22nd TCS were reassigned into the newly designated 374th TCG Group. On 29 January 1943 both squadrons moved to Wards 'drome, Port Moresby.

Allocated squadron numbers 26 to 50 during its time in the SWPA, the 22nd TCS lost eight transports to operational accidents in the theatre, including one missing after departing Essendon on 14 July 1942. In addition to the C-47s profiled here, the following names also appeared on squadron C-47s: *Guinea Gopher*, *L'il Abner*, *Overtime*, *Pel* and *Smokey Joe*.

The 22nd TCS adopted a logo featuring a pack-laden donkey climbing a steep hill and this was painted on several of its C-56s.

The subject of Profile 15, C-47 serial #41-18645 Linda Ann, airborne over the Australian coast. It has been photographed from another C-47.

21st TCS

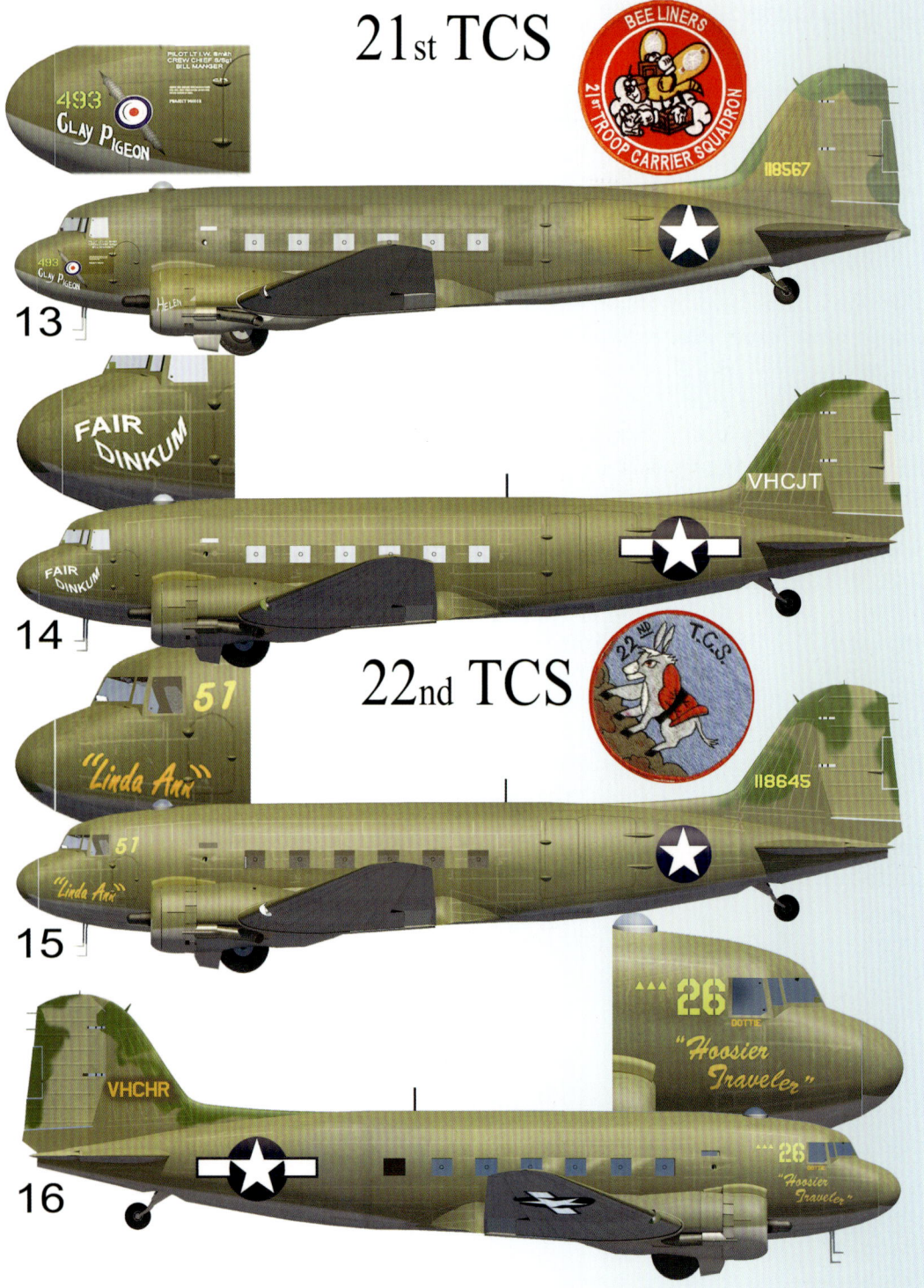

BEE LINERS
21st TROOP CARRIER SQUADRON

118567

13

14

FAIR DINKUM

VHCJT

22nd TCS

22ND T.C.S.

118645

15

51

"Linda Ann"

26
DOTTIE
"Hoosier Traveler"

26
"Hoosier Traveler"

VHCHR

16

Profile 13 – C-47 serial #41-18567, squadron number 493, *Clay Pigeon*

Clay Pigeon was originally assigned to the 41st TCS on 13 December 42 and then flown to Australia. It was handed over to the 21st TCS at Port Moresby on 27 January 1943. This transport continued to carry its original stencilled stateside squadron number 493 until around May 1943 when it was assigned an unknown squadron number. The aircraft ended the war serving with the 21st TCS. It suffered a repairable operational accident on 14 September 1944, after which it returned to the US on 21 January 1945.

Profile 14 – C-47A-15 serial #42-92802, VHCJT, *Fair Dinkum*

Assigned into the Fifth Air Force on 9 April 1944, this transport at first served with DAT whilst assigned to the 21st TCS. On 14 December 1944 it was badly damaged by severe turbulence while flying between Sorido and Morotai. It was then grounded pending extensive repairs. The aircraft is illustrated as it appeared at Goodenough Island, when it delivered supplies there in mid-1944. Note the white trim tab.

Profile 15 – C-47 serial #41-18645, squadron number 51, *Linda Ann*

This transport was first assigned to the 22nd TCS on 12 November 1942. It arrived at Ward's 'drome on 12 January 1943 where, exceptionally, it was given squadron number 51 which lay just outside the allocated squadron range of 26 to 50. It later served with DAT with which it was allocated radio callsign VHCGB.

Profile 16 – C-47A-60 serial #43-30742, squadron number 26, VHCHR, *Hoosier Traveler*

Issued to the 22nd TCS on 7 October 1943 and allocated squadron number 26 and DAT callsign VHCHR, this transport was named *Hoosier Traveler* (sic) by pilot Lieutenant William Crecelius, from Indiana. The stencilled name *Dottie* also appeared underneath the starboard cockpit window. By way of explanation, Hoosier is a colloquial term used to describe Indiana residents. Its origin remains unclear, but Hoosier had attained widespread use by the 1840s, having been popularized by a poem of the time *The Hoosier's Nest*.

This aircraft underwent a mysterious loss on the morning of 19 December 1943, when it disintegrated about 30 miles north of Rockhampton in Queensland. It was *en route* from Garbutt airfield to Archerfield via Rockhampton, ferrying US and Australian personnel for leave or administrative reasons. At 0945 hours a young Australian Volunteer Air Observer Corps eyewitness saw the aircraft explode and fall to earth. No cause of crash was ever determined, however, 31 personnel lost their lives in the crash, including three US Army nurses. It remains Australia's second-worst air disaster at the time of publication.

The crew of C-47 serial #41-38660 Gremlin, the subject of Profile 18, at Wards 'drome.

A line-up of 33rd TCS transports about to depart Port Moresby's Seven-Mile 'drome.

CHAPTER 6
33rd Troop Carrier Squadron

On 18 October 1942 the 33rd TCS departed Hamilton Field with thirteen brand new C-47s, bound for the SWPA. Upon their arrival at Canton Island via Hawaii three days later, ten were sidelined to assist with an aerial search for Captain Eddie Rickenbacker and his crew missing in a ditched B-17E. Following a search which lasted two days, all C-47s had proceeded to New Caledonia by the late afternoon of 25 October. However, when they arrived six aircraft and their crews were commandeered by the USN to ferry personnel and supplies to Guadalcanal via the New Hebrides, using Tontouta as the main base.

During the delivery stopover at Nadi, Fiji, and without relevant personnel paperwork, several pilots with the rank of staff sergeant were detained. With local American authorities unfamiliar with the fact that non-officers had qualified as pilots, their alleged crime was stealing a C-47 to fly AWOL to Australia. Next morning, a USMC R4D flew in from Tontouta, the pilot of which requested they all board the squadron's C-47 *Man-o-War* and follow the R4D back to New Caledonia. There the "flying sergeants" would be detained pending trial. The two aircraft headed for New Caledonia, however on approaching the island in poor visibility, *Man-o-War* broke off and set course for Plaine des Gaiacs. Here other 33rd TCS C-47s had already landed, and the matter was quickly cleared up.

The next morning *Man-o-War* journeyed to Brisbane during an uneventful six-and-a-half-hour trip. On 29 October it was followed by the other six C-47s not held at New Caledonia for SOPAC duty. They headed northwards via Cairns for Ward's 'drome, Port Moresby, where they arrived on 2 November.

Meanwhile for those which had remained under SOPAC control, on 5 November the 33rd TCS lost its first C-47 when *Pack Rat* burst into flame after take-off from Henderson Field, having been struck by Japanese ground fire. Then, on 9 November, *Full House* was loaded with cases of hand grenades and was demolished following a take-off crash at Espiritu Santo.

The 33rd TCS was allocated squadron numbers 76 to 99 upon arrival at Ward's 'drome, and a high tempo of operational missions quickly began. On 28 December the ground echelon arrived at Port Moresby. This was the same day that Captain Eugene Jackson was appointed squadron commander. He served until replaced by Captain George Wamsley on 15 October 1943. During this period the 33rd TCS lost six C-47s to operational causes and four to enemy activity.

One of the operational losses showcases a classic case of mixing aircraft and crew assignments. On 10 October 1943 a 65th TCS crew captained by John Hutchinson was operating the 33rd TCS's C-47 serial 41-38761 when he force-landed it six miles from Nadzab. Such mixing of crews and planes from different squadrons was quite common as operational demands required.

In addition to the C-47s profiled here, the following were names also given to other 33rd TCS

33rd TCS

86
THE Jayhawk 2nd

86
THE Jayhawk

138665

17

85
"GREMLIN"

85
"GREMLIN"

138660

18

"LITTLE SKUNK FACE"

"LITTLE SKUNK FACE"

118601

19

642 EARLY DELIVERY

642 EARLY DELIVERY

138658

20

C-47s: *Burma Girl, Chattanooga Choo Choo, The Cherokee, Daffy, Eager Beaver, The Flying Dutchman, Full House, Hell's Angel, The Jayhawk, Mari Beth, Miss Carriage, The Mohican, The Navajo, Pack Rat, Pueblo, Pushy Cat, Queenie, The Shack Trooper, Shady Lady* and *Shanghai Lil.* The 33rd TCS adopted a logo of a turtle carrying a back-pack over clouds.

Profile 17 – C-47 serial #41-38665, squadron number 86, *The Jayhawk 2nd*

This transport was first issued to the 46th TCS in the US on 17 November 1942 which named the aircraft. It was flown across the Pacific by this squadron, but it was transferred to the 33rd TCS on 24 January 1943 at Port Moresby and allocated squadron number 86. The aircraft was destroyed in an accident in the Philippines on 31 December 1945.

Profile 18 –C-47 serial #41-38660, squadron number 85, *Gremlin*

This transport was another issued to the 46th TCS in the US on 17 November 1942 which named it *Gremlin*. This squadron flew it to Australia before handing it over to the 33rd TCS upon arrival at Wards 'drome on 24 January 1943. It saw out its war service with the 33rd TCS, before returning to the US on 26 January 1945.

Profile 19 –C-47 serial #41-18601, *Little Skunk Face*

This C-47 was transferred to the 33rd TCS on 18 September 1942 with which it was subsequently named. It one of the seven transports detained in New Caledonia for SOPAC duty on 25 October 1942 before finally proceeding to New Guinea on 28 November. There it was later transferred to the 41st TCS on 2 February 1943 which renamed it *South Wind Yankee Flyer*. From August 1943 it served with DAT firstly as VHCCS and later as VHCFX and with the name *South Wind*. The C-47 returned to the US on 20 March 1945 and is portrayed as it served in New Caledonia without a squadron number.

Profile 20 –C-47 serial #41-38658, squadron number 642, *Early Delivery*

Originally assigned to the 46th TCS as squadron number 642 and named *Early Delivery*, this C-47 was transferred to the 33rd TCS upon arrival at Wards 'drome on 24 January 1943. However, it was never given a new squadron number in the 76 to 99 range as it was lost before the system was implemented. On 6 February 1943 *Early Delivery* was among a flight of four 33rd TCS C-47s which entered a holding pattern over Wau, waiting to deliver supplies, when Japanese fighters appeared. Lieutenant Robert Schwensen made an emergency landing on Wau's uphill grass airfield in this aircraft, following by the other three C-47s. Bombs started dropping so the C-47s chose to immediately depart, however *Early Delivery* was shot down by a JAAF 11th *Sentai* Ki-43-I *Hayabusa* as it climbed out from Wau. It crashed into hillside jungle near the village of Wandumi, with the wreckage not found until 1988. The aircraft is profiled as it appeared on the day it was handed over to the 33rd TCS at Wards 'drome, still in its original 46th TCS markings.

Profile 21 – C-47 serial #41-18665, squadron number 80, *Queenie*

Originally assigned to the 39th TCS which named it *Queenie*, this C-47 was transferred to the 33rd TCS on arrival at Wards 'drome on 24 January 1943. It served out its days in the theatre as squadron number 80, before returning to the US on 4 February 1945.

Profile 22 – C-47 serial #41-18602, *Man-O-War*

Delivered to the USAAF on 16 September 1942, pilot Staff Sergeant Claude Patterson named this new C-47 *Man-O-War* when he received it a few days later. The next day he flew to Mobile, Alabama, to acquire a set of Jeep loading ramps, thence on to San Antonio, Texas, where the crew joined the other twelve 33rd TCS transports for the trans-Pacific delivery flight to Australia. *Man-O-War* was so busy during the Buna/Gona campaign that DAT lost track of its movements for an entire week, and mistakenly reported the aircraft as missing in action. It was transferred to the 41st TCS in Townsville on 2 February 1943 which renamed it *Wabash Cannonball* with the DAT callsign VHCCT. It dropped paratroopers at Nadzab as part of C Flight on 5 September 1943, and later operated with DAT in 1944 as VHCFR. It returned to the US on 12 December 1945.

The crew of C-47 serial #41-38665 The Jayhawk 2nd pose in front of their aircraft at Wards' drome. It is the subject of Profile 17.

C-47 serial #41-18601 Little Skunk Face, the subject of Profile 19, seen during a liaison flight to Sydney.

The subject of Profile 24, C-47 serial #41-18659 Miss Carriage, in a revetment at Wards 'drome, Port Moresby.

Later in the war the 317th TCG named itself The Jungle Skippers as seen painted on one of its C-47s in the Philippines on 22 June 1945.

CHAPTER 7
39th Troop Carrier Squadron

The 39th TS was formed at Duncan Field outside San Antonio, Texas, in early 1942 and is one of the few wartime USAAF transport squadrons still active at the time of publication. Currently operating as the USAF 39th Airlift Squadron equipped with C-130J Super Hercules transports, it still uses the same wartime squadron insignia. This is an emblem of a flying wagon train and was created on 27 October 1942. It subsequently appeared on nearly all of the squadron's C-47s.

After formation in early 1942 the squadron spent some months training with DC-3s requisitioned from US airlines. The 39th TS was redesignated as the 39th TCS on 4 July 1942. At exactly midnight on 5 January 1943 the commander of the 39th TCS, Lieutenant Joseph Ford, departed Hamilton Field, California, for Australia followed by the rest of the squadron. It was the first of four squadrons in the 317th TCG to leave the US, closely followed by the 40th TCS the next day, the 41st TCS the next night and the 46th TCS three days later. Most pilots had barely accumulated 500 hours total flying time, and for most navigators this was their first flight outside the training environment.

Meanwhile over in Australia in late December 1942, 21 Dutch aircrew had been designated a separate "transport unit" which was seconded to the Fifth Air Force and then attached as a separate flight to the 39th TCS when it arrived in Australia at Camp Muckley, about a mile from Archerfield airfield just to the south of Brisbane (see also Chapter 28 for more detail on this Dutch flight). Here the squadron also trained on CG-4 gliders, although these were never used in combat in New Guinea.

For its first forward deployment the 39th TCS was first based at Seven-Mile, Port Moresby, and Ford, now promoted to Captain, led the squadron on its first combat mission on 30 January 1943 to deliver supplies to Australian soldiers at Wau. After the Battle of Wau, the squadron rejoined its ground echelon which meanwhile had first arrived at Brisbane aboard the SS *Maui* and then moved to Garbutt Field, Townsville. By April 1943 the 39th TCS was assigned to DAT with which it was also operating six C-60s and three C-39s.

After the unit transferred to Finschhafen in early 1944, Ford (who had been promoted again to Major), lost his life on 8 May 1944 in a fiery take off crash on a flight destined for Gusap. During the climb out of Finschhafen the starboard engine seized, drifting the aircraft starboard while the crew apparently attempted to feather the engine. The slow transport hit trees and Major Ford died late that afternoon from injuries received.

During its time in the SWPA the 39th TCS was allocated squadron numbers 1 to 25 and lost nine C-47s to operational accidents and one to an enemy air raid. Names/nose art used in the theatre include *Barbara B*, *Breezy Ann*, *Bubbles*, *Chattanooga Choo Choo*, *Cleo*, *Duchess*, *Frigid Midget*, *Usta Olfingwa*, *Gearann*, *Gremlin Special*, *Guinea Gopher*, *Hairless Joe*, *Hell's Bells*, *The Lonesome Angel*, *Louise*, *Loveable Lou*, *Margie*, *Miss Carriage*, *Oki*, *Ready Willing And Able*, *Slut II*, *We Dood It* and *Joanne*.

39th TCS

Profile 23 – C-47 serial #42-23955, squadron number 16, *The Lonesome Angel*

This transport later served with the DAT as VHCHI.

Profile 24 – C-47 serial #41-18659, squadron number 643, *Miss Carriage*

This transport was received from the 33rd TCS which gave it squadron number 84, painted over in this profile. The number 643 applied in the US remained on the airframe throughout its SWPA service.

The subject of Profile 23, C-47 serial #42-23955 The Lonesome Angel, at Wards 'drome, Port Moresby.

A Kodachrome photo of Bubbles in Sydney just after it was transferred out of the 39th TCS.

Beach Comber was serial number 43-15478 assigned to the 40th TCS as squadron number 31 on 11 May 1944. It is seen here at Townsville where it had wing repairs following an accident on 28 August 1944. Note the number 31 visible behind the cockpit which has been painted over during repairs.

CHAPTER 8
40th Troop Carrier Squadron

The 40th TCS was the second of the four comprising the 317th TCG to leave Hamilton Field, California, for Australia. Like most pilots in the group, most of the 40th TCS pilots had barely accumulated 500 hours total flying time. However, the squadron entered the Battle for Wau only a few days after arriving in New Guinea.

On 18 January 1943 the 40th TCS suffered its first loss, when Lieutenant Robert Sams flared too high on approach to Wau, likely as the result of an optical illusion when misjudging the steep incline of the field. The aircraft clipped a tree and cartwheeled into the jungle short of the runway, where it was destroyed by fire. The radio operator, Private Edward Johnston, was the only survivor. Despite severe injuries he eventually recovered and returned to duty with the squadron.

By April 1943 the 40th TCS was assigned to DAT and was operating nine C-47s. On 7 August 1943 the squadron suffered its biggest loss when *Eager Beaver* crashed into Cleveland Bay off Townsville at 0520 shortly after take-off, bound for Archerfield in Brisbane. The aircraft had 27 aboard mainly aircrew from the 345th BG *en route* to Sydney for leave while their B-25Ds were being modified as low-level strafers at the Townsville Air Depot. The loss set back the 345th BG considerably as many air trained crewmembers - turret gunners, radio operators and flight engineers - had to be replaced at short notice.

On 3 February 1944 an exceptional loss occurred when *Missouri Mule* was fired upon by Allied vessels off Finschhafen in the belief that it was a Japanese bomber. The aircraft was hit and set on fire. It was ditched and the crew was rescued, although most were suffering from burns.

In the SWPA the 40th TCS was allocated squadron numbers 26 to 50, and it lost five aircraft to operational accidents and one to friendly fire as described above. In addition to the four C-47s profiled here, the following names also appeared on squadron C-47s: *Aeolus, Barbara Ann, Betty El, Blonde Baby, Blue Grass Baby, Cheeky Charley, Duchess, Dutchess, Eager Beaver, Estelle, Flamingo, The Georgia Peach, Golden State Arrow, Hot Box, Lone, Kingfish, The Last Straw, Mary, Miscellaneous, Mississippi, Missouri Mule, Open Date, Rusty, Shamrock, We Dood It, Mokano, Sleepy Saloon, Stepinfetchit, Stuff* and *Yank's Delight*.

40th TCS

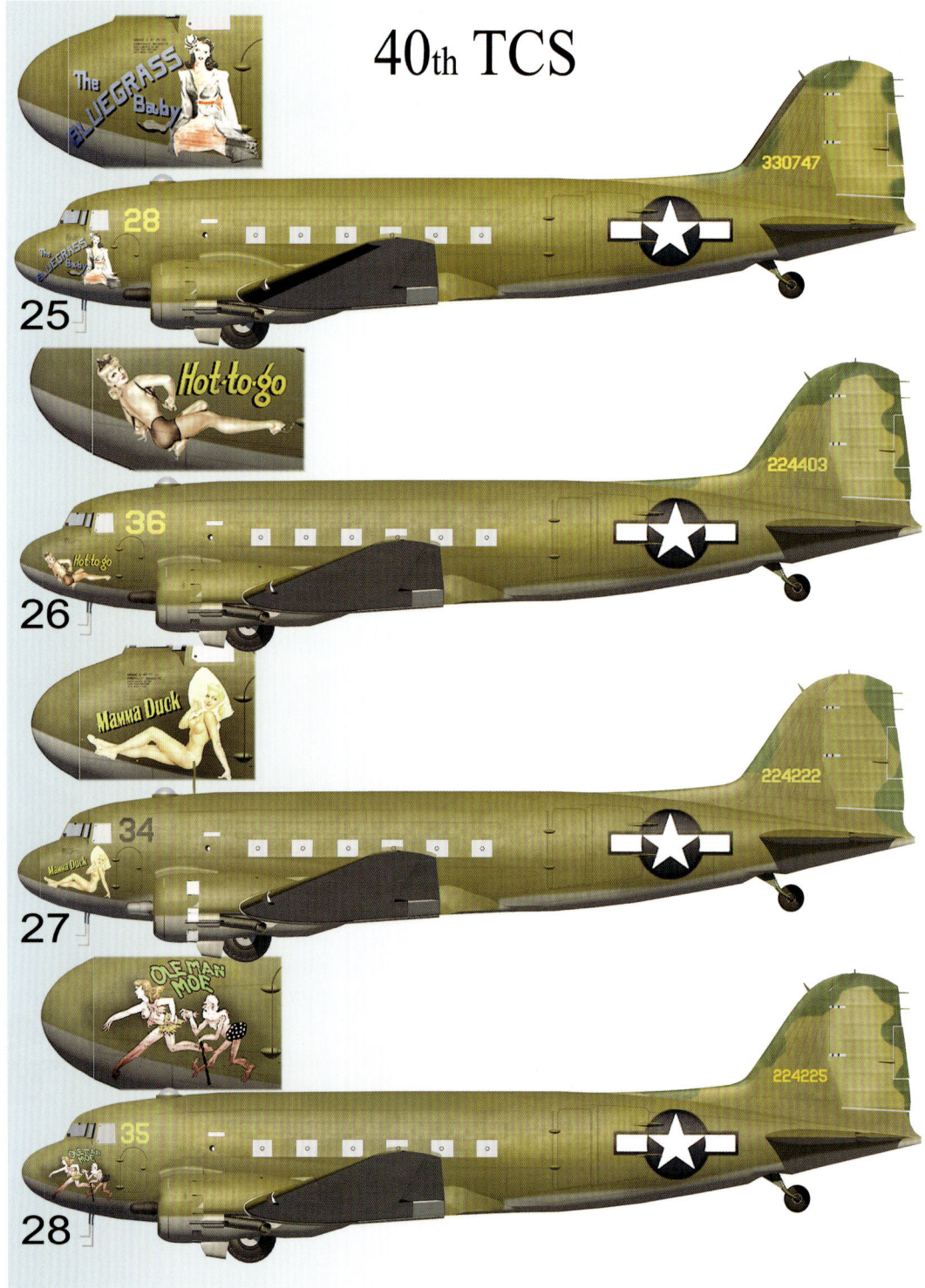

25

26

27

28

Profile 25 – C-47 serial #43-30747, squadron number 28, *The Bluegrass Baby*

The aircraft is profiled as it appeared at Finschhafen in early 1944.

Profile 26 – C-47 serial #42-24403, squadron number 36, *Hot-to-go*

This transport was assigned to the 40th TCS on 30 October 1944. It survived the war and was written off in the US in 1947.

Profile 27 – C-47 serial #42-24222, squadron number 34, *Mamma Duck*

The aircraft is profiled as it appeared at Finschhafen in early 1944. It was later destroyed in a landing accident at Noemfoor on 20 November 1944.

Profile 28 – C-47 serial #42-24225, squadron number 35, *Ole Man Moe*

The aircraft is profiled as it appeared at Nadzab in late 1943.

Flamingo at an unidentified coastal airfield on New Guinea. This transport was previously named Pel and later served with DAT as VHCCQ.

C-47 serial 41-18539 first served with the 6th TCS, arriving in Australia on 9 October 1942. It was transferred to the 40th TCS on 17 January 1943. Named Sleepy Saloon, it was allocated the DAT callsign VHCCL. After the war it went to the US civil registry, then later served with the Argentine Navy as CTA-15. It became the first Argentine aircraft to land at the South Pole.

Blonde Baby was serial number 42-23588 received by the 40th TCS on 8 June 1943 and given squadron number 28. It was later allocated the DAT callsign VHCGP and was transferred to the 39th TCS in November 1943 where it was allocated squadron number 18. It is seen here at Townsville being repaired, between these two assignments.

Hot Box also served with the 40th TCS although its serial number is unknown. It is seen here in the Philippines after being reassigned to another unit and with its squadron number removed.

C-47 serial number 42-23582 Honeymoon Express, the subject of Profile 31, at Port Moresby in the lead-up to the Nadzab paratrooper drop of 5 September 1943.

CHAPTER 9
41ˢᵗ Troop Carrier Squadron

The 41ˢᵗ TCS was the third of the four squadrons which comprised the 317ᵗʰ TCG that commenced leaving Hamilton Field, California, for the SWPA on 7 January 1943. Commanded by Captain Herbert Waldman, a 24-year-old statistician from Long Island, New York, its pilots were fresh out of training school each with around 500 hours flying time. The 41ˢᵗ TCS entered the Battle for Wau in late January 1943 shortly after arriving in New Guinea. After this the squadron was transferred to DAT, and by April 1943 it was operating nine C-47's.

The day after the Nadzab drop in which it participated, on 6 September 1943 a dozen of the squadron's C-47s hauled cargo to Tsili Tsili in connection with the air movement of the 7ᵗʰ Australian Division. Alongside many other transport units, the 41ˢᵗ TCS then flew continuous shuttles to help build Nadzab's new airfields until 19 September. On 20 September, all of the squadron's C-47s returned to Townsville in a single formation to transport the 317ᵗʰ TCG ground echelon to Ward's 'drome, Port Moresby.

The 41ˢᵗ TCS C-47s were assigned squadron numbers 51 to 75, and it adopted the logo of a walking mother cat carrying its kitten. The squadron lost three aircraft in the theatre, all to operational causes. Many of the squadron's transports adopted a train theme for their names, examples include *The Hiawatha* (a model of locomotive operated by the Pacific Railroad), *Silver Meteor* (1939 first streamliner diesel), *The Texas Special*, *The Pacemaker* (New York Central Railroad), *Fireball Mail*, *Golden State Flyer*, *The Rocket* (named after *Stephenson's Rocket*), *The Zephyr* (1934 diesel), *The Pathfinder* (New York Central Railways), *Honeymoon Express* and *The Pan American*. After the war, the former squadron commanding officer Herbert Waldman rose to the rank of Major-General and later became Chief of Staff for Military Airlift Command.

The 41ˢᵗ TCS flight line at Dobodura in early 1943, with Miss Ohio, the subject of Profile 36, at the front.

41st TCS

29

30

31

32

Profile 29 – C-47A serial #43-15465, squadron number 63, *Hells Bells II*

This transport was assigned to the 41ˢᵗ TCS around April 1944. It delivered entertainer Bob Hope and his entourage to Hollandia to conduct a United Services Organization show there in August 1944. The *"II"* from *Hells Bells II* in this profile is barely discernible, but it is at the bottom left-hand corner of the forward door.

Profile 30 – C-47 serial #41-18568, squadron number 54, *Dumbo Junior*

This aircraft was delivered by the 6ᵗʰ TCS to Australia on 21 January 1943, and later operated as VHCCM. It was transferred to the 41ˢᵗ TCS in October 1943 where it was briefly operated as *Dumbo Junior* (as profiled) before it was renamed *The Broadway Ltd*. The aircraft was lost on 1 April 1944 when it crashed five miles southwest of Dobodura #7 strip. This was due to a starboard engine failure on take-off, and although Lieutenant Henry Ellis feathered the engine, the transport hit trees while turning back for the strip.

Profile 31 – C-47 serial #42-23582, squadron number 62, *Honeymoon Express*

This transport was named shortly after it was received by the 41ˢᵗ TCS in June 1943. It became briefly famous when Colonel Paul Prentiss, the commander of the 54ᵗʰ Troop Carrier Wing, flew as co-pilot in the transport when it was the lead C-47 for the Nadzab air drop of 5 September 1943. The aircraft is profiled just before it was transferred to DAT with the radio callsign VHCGQ.

Profile 32 – C-47A serial #42-23856, squadron number 64, *Polly*

Assigned to the 41ˢᵗ TCS on 27 August 1943, this participant in the 5 September 1943 Nadzab air drop was out of service for several weeks after colliding with another C-47 near Dumpu on 13 June 1944. It suffered a second collision at Biak just after it was repaired and was later written off at Tacloban in the Philippines.

41st TCS

Profile 33 – C-47A serial #42-93497, squadron number 55, *Cajun Coonass*

Delivered to the 41st TCS in June 1944, this transport was damaged by an air raid in the Philippines in 1945.

Profile 34 – C-47 serial #41-18648, squadron number 610, *Star Duster*

This transport was named in the US while at Hamilton Field, where it was also given the squadron number 610. It is profiled as it appeared at Wards 'drome before it was allocated an unknown SWPA squadron number in the range 51 to 75. By mid-1943 it was serving with DAT with the callsign VHCGC and was operated by the 21st TCS. On 21 November 1943, it was lost in a freak accident on the final leg of a passenger flight from Wards 'drome to Archerfield via Cairns, Townsville and Rockhampton, carrying ten passengers and a crew of four. At Townsville two vehicle engines along with spares were also loaded aboard. After departing Rockhampton just before midday, the aircraft headed directly for Archerfield then disappeared without trace. The wreckage was located in 1948 in a steep gulley in the inland Monto area, about halfway between Rockhampton and Archerfield.

The subsequent investigation revealed that the aircraft had disintegrated in mid-air, possibly due to severe turbulence which stressed an overloaded airframe. The entire starboard wing with the engine nacelle was located about a kilometre from the fuselage, while the starboard engine was located 400 metres from the fuselage.

Profile 35 – C-47 serial #42-23423, squadron number 59, *Miss America*

This C-47 was delivered to the 41st TCS in June 1943 from the RAAF where it had been A65-5 with the DAT radio callsign VHCTE. It saw out the war and then remained in New Guinea for civilian service. It served Guinea Airways in 1955 which named it *Kokoda*, before later service with Ansett Express, Airlines of South Australia, Airlines of NSW and Ansett-ANA. It was finally scrapped on 4 November 1969 at Essendon airport, near Melbourne.

Profile 36 – C-47 serial #41-38661, squadron number 58, *Miss Ohio*

Delivered to the 41st TCS on 14 December 1942, *Miss Ohio* saw later service with DAT as VHCJB. It survived the war to briefly fly with TWA on the US civil registry as NC86567 in 1947, however, its final fate is unknown.

C-47 serial number 41-19472 *Battling Bishop* was assigned to the 41st TCS after having first served the 21st TCS.

C-47 serial number 42-23423 *Miss America*, the subject of Profile 35, at Finschhafen.

This C-47, serial number 42-23583, was first named *The Pathfinder* when received by the 41st TCS on 30 May 1943 but was later renamed *Flamingo* as seen here at Wards 'drome. It also served with DAT as VHCGJ and later with the Fifth Air Force Service Command at the end of the war.

C-47A serial number 42-23856 *Polly*, the subject of Profile 32, at Wards 'drome.

On 21 June 1943 C-47A 42-23659 was assigned into the 46th TCS and given the DAT callsign VHCGX, with this nose art and the name *The Amazon* applied at Wards 'drome. It arrived with the squadron number 438 applied in the US (as pictured), which was replaced by squadron number 85. On 1 December 1943 with four crew onboard it departed Garbutt Field flown by First Lieutenant James Nollkamper bound for Wards 'drome, Port Moresby on a cargo run. Both engines failed at 4,000 feet while west of and short of Port Moresby, forcing the transport to force-land in a swamp on the bank of the Vanapa River. The crew were unhurt and rescued the following day by an Australian launch which sailed up the river to locate them.

At time of publication the wreckage of The Amazon remains in situ, as shown by this recent photograph. (courtesy Justin Taylan/ pacificwrecks.com).

CHAPTER 10
46th Troop Carrier Squadron

The 46th TCS was the last of the four comprising the 317th TCG to leave Hamilton Field, California, for the SWPA in early January 1943. The 46th TCS departed on 8 January 1943 led by Captain James Evans. In common with the other squadrons, most pilots had barely accumulated 500 hours of total flying time. The squadron adopted a logo of an aircraft carrying a relaxed soldier in a hammock.

The 46th TCS entered the Battle for Wau only a few days after arriving in New Guinea. The uphill Wau airfield was particularly treacherous in wet weather. On 30 January 1943 a 21st TCS C-47 barrelled in between two other parked C-47s, and a spinning propeller severed the left leg below the knee of 46th TCS pilot Flying Officer William Teague, making him the squadron's only casualty in the battle for Wau.

After the Battle of Wau, the entire 317th TCG returned to Australia where it joined the ground echelon which had arrived in Australia on the SS *Maui* and setup camp at Garbutt Field, Townsville. The 46th TCS concentrated on three major operations from February through August 1943, however not before reluctantly swapping its entire inventory with the 374th TCG. This was done as the latter group needed the newer aircraft for front-line service while the 46th TCS operated cargo runs with DAT. In April 1943 DAT records indicate that the squadron, alongside the hand-me-down C-47s, was also operating five C-49s, two LB-30s, two B-17Cs and several DC-3 converted airliners.

The 46th TCS commenced moving troops and materiel from Air Transport Command Australian terminals to Port Moresby or Milne Bay. It also delivered this materiel to forward units. During this period the squadron seconded crews to the 374th TCG to gain front-line experience, and it also trained with the 503rd Parachute Infantry Regiment by practising paratrooper drops near Cairns. Captain James Evans led the 46th TCS as C Flight during the Nadzab paratrooper drop of 5 September 1943 under the call sign "Vesper C".

46th TCS

37

38

39

40

Profile 37 – C-47 serial #41-18651, squadron number 76, *Chattanooga Choo-Choo*,

This transport commenced service with the 46th TCS in January 1943 but on 12 May 1943 it was seconded to the 21st TCS with a mixed 41st TCS crew. It departed Dobodura flown by Second Lieutenant Lorenzo Gower transporting cargo and passengers bound for Wards 'drome, Port Moresby, but never arrived. Confusion has existed over the years as to the aircraft's squadron assignment due to the mixed crew, followed by a wrong serial number entry in the Missing Air Crew Report. Those aboard included four crew and six passengers among which were three civilians and a Papuan cook. The aircraft crashed into mountains in the Kokoda region near Abuari, killing all aboard.

Profile 38 – C-47 serial #42-23489, squadron number 76, *Miss Priss*

This transport was named as soon as it was received by the 46th TCS on 12 May 1943. It survived the war and returned to the US post-war. Its squadron number 76 replaced that worn by *Chattanooga Choo-Choo* following its loss, also on 12 May 1943 (see Profile 37 above). Note the red surround on the US insignia.

Profile 39 – C-47 serial #42-23716, squadron number 80, *What's Cookin' 2nd*

This transport entered service with the 46th TCS in late May 1943 and survived the war after which it saw service with the Argentine Navy. At time of publication, it was under static restoration in Argentina.

Profile 40 – C-47 serial #42-23656, squadron number 86, *The Rube*

Named *The Rube* as soon as it entered service with the 46th TCS on 19 June 1943, this transport also operated with DAT as both VHCGU and VHPAC. After the war it was transferred to the Netherlands Navy as Q-3 from where it was transferred to the Indonesian Air Force on 15 December 1950.

C-47 serial number 41-18651 Chattanooga Choo-Choo, the subject of Profile 37, at Bena Bena in the New Guinea highlands, prior to its loss on 12 May 1943.

C-47 serial number 42-23716 What's Cookin' 2nd, the subject of Profile 39, has an engine problem addressed at Wards 'drome.

The crew of Miss Priss, the subject of Profile 38, pose with their aircraft at Seven-Mile, Port Moresby. A B-17E is in the background.

Lazy lady was assigned into the 46th TCS in June 1943. It transported Hollywood actor Gary Cooper and party to Dobodura to put on a United Service Organization show in November 1943 as seen here. It was returned to the US after the war and was finally lost in 1947 while in civilian service when it crashed near Caracas, Venezuela.

Texas Honey, the subject of Profile 41, shortly after making its famous landing at Hollandia on 23 April 1944. It has been caught by an official US Army photographer as it turns around prior to shutting down its engines. The wreckage of a Japanese Army Air Force Hayabusa fighter lies in the foreground.

The nose art on Texas Honey, the subject of Profile 41, is seen at Garbutt whilst the aircraft was undergoing repairs.

CHAPTER 11
55th Troop Carrier Squadron

The 55th TCS was one of four squadrons activated within the 375th TCG at Bowman Field, Kentucky, on 18 November 1942. It commenced operations from Port Moresby on 15 July 1943, and moved its headquarters to Dobodura 19 August 1943, then back to Port Moresby (Wards 'drome) on 22 December 1943 and then on to Nadzab on 22 April 1944. The squadron left the SWPA when it moved to Biak on 1 October 1944.

The 55th TCS's most famous aircraft was *Texas Honey* (see Profile 41), the first Allied aircraft to land at newly captured Hollandia. Following its capture on 22 April 1944, General Ennis Whitehead needed to get urgently needed documents to Hollandia as soon as possible and tasked his assistant Colonel AJ Beck to make it happen. Beck sought out the commander of the 375th TCG who assigned *Texas Honey* the task. Beck carried the documents himself on the flight which departed Nadzab the following day. The landing was a close-run thing as described by the pilot:

> It came to a choice between Cyclops, Sentani or a deliberate crashlanding. Cyclops was the best bet, but it would have to be a shorter landing than either of us had ever made ... we hit short of the runway on the overrun. I thought the gear would come up through the wings ... we missed a couple of bomb craters in the runway and stopped just short of a trench.

Texas Honey returned to Nadzab the following morning with 23 stretcher cases. The 55th TCS only lost two C-47s to operational causes: a ditching off Owi in September 1944 and an unusual loss north of Essendon airfield on 10 August 1944. The transport had departed Brisbane with four crew and 21 USAAF passengers, however, a weather front north of Newcastle diverted the aircraft to Melbourne. The pilot made a forced landing at Broadmeadows, to the north of Essendon airport, and skidded out of control. It careered into a wide stormwater drain and was written off, although only two of the crew suffered minor injuries.

55th TCS C-47A serial number 42-23876 featured colourful nose art painted by artist Al Merkling, Note the squadron number 106 also appears under the nose.

55th TCS

Profile 41 – C-47A serial #42-32805, squadron number 111, *Texas Honey*

This transport was assigned into the 55th TCS as squadron number 111 at Brisbane on 11 July 1943. It is profiled as it appeared during its landmark landing at Hollandia on 23 April 1944. Following these operations, it survived the war and it was returned to the US on 23 August 1945.

Profile 42 – C-47 serial #41-38681, squadron number 101, *Bobcat*

This C-47 was received by the 55th TCS at Brisbane on 8 February 1943 with the US assigned squadron number 677. Subsequently allocated 55th TCS squadron number 101, it was lost when it experienced problems *en route* from Owi to Finschhafen on 24 September 1944 and was ditched.

55th TCS C-47 serial number 42-24223 at Cape Gloucester in front of other transports and with 35th Fighter Group P-40Ns parked in the background. This C-47 was destroyed in an accident in the Philippines in July 1945.

55th TCS C-47 squadron number 103 lies wrecked after a forced landing near Essendon airport on 10 August 1944. It was one of only two aircraft lost by the squadron during its time in the SWPA.

C-47A serial number 42-24396, the subject of Profile 43, is salvaged shortly after its ditching off Magnetic Island.

The nose art on Katie which served the 56ᵗʰ TCS as squadron number 138. Its serial number is unknown.

Shakes All Over with its crew at Munda in the Solomons in April 1944, when the squadron fell under Thirteenth Air Force command for several weeks to assist with provisioning the base.

CHAPTER 13
57th Troop Carrier Squadron "King's Men"

The 57th TCS was activated on 18 November 1942 at Bowman Field, Kentucky, and assigned to the 375th TCG alongside the 55th, 56th and 58th TCS. The squadron's first commander was Captain Lucian Powell, a former airline pilot with over 6,000 hours flying time. The unit received most of its key personnel from the 30th TCS then operating at Sedalia, Missouri. Powell was replaced by Lieutenant Benjamin King in March 1943. It was this change of command that coined the 57th TCS's nickname of "King's Men". However, this is not associated with the adopted squadron logo that depicts a native North American woman carrying an armed solider on her back.

In preparation for overseas deployment the new squadron's training regime included transporting cargo and personnel, airborne drops and towing gliders. On 2 June 1943, it moved to Baer Field, Indiana, where it received more equipment and underwent final training before its deployment to the SWPA. On 14 June the first C-47s departed the US, arriving at Brisbane on 23 June before continuing to Port Moresby three days later. The 57th TCS moved to Dobodura on 2 August and two days later flew its first combat delivery being a supply drop to Salamaua. On 2 September nine C-47s dropped 90,000 pounds of ammunition, food and equipment to Allied troops. After 16 October the air echelon returned to Port Moresby, with support crews moving back there on 20 December 1943.

In addition to C-47s, the 57th TCS operated B-17F Fortresses in the transport role from February to May 1944. From 8 April 1944 for several weeks the squadron briefly fell under Thirteenth Air Force command when it delivered personnel and equipment from Guadalcanal to Munda. The unit supported the invasion of Noemfoor Island on 2 July 1944 and then left the SWPA when it moved to Biak on 23 September 1944.

The worst operational loss in New Guinea for the 57th TCS occurred on 9 July 1944 when *Shakes All Over* departed Saidor Airfield flown by Lieutenant Marvin Davis. Its four crew were delivering two passengers and cargo to Nadzab, including about two dozen Browning 0.50-inch calibre M2 machine guns requiring service, however it failed to arrive. The C-47 crashed into mountains at 7,400 feet altitude in the vicinity of the Saidor Gap in the Finesterre Ranges, from where the crew remains were recovered in 1979.

Note that *Shakes All Over* had the squadron number X163 as early as April 1944. As explained in Chapter 2, a W or X prefix was added to the squadron number during the Philippine campaign in 1944-45 to differentiate between transports and other types of aircraft. The use of the X prefix by *Shakes All Over* in New Guinea during the first half of 1944 likely indicates an earlier experimentation with this system.

The 57th TCS operated a C-47 it named *Kings Cross Shuttle*, one of the squadron's originals which had arrived in Australia on 26 June 1943. It arguably became the Fifth Air Force aircraft with the most lives, after experiencing a string of accidents and repairs: it was struck by a Jeep at Dobodura in October 1943, and after repairs was then hit by another taxiing C-47. It was damaged whilst taxiing at 17-Mile (Port Moresby) on 12 March 1944, then again whilst taxiing at Finschhafen on 6 April 1944. After repairs it was damaged taxiing at Wards 'drome eight days later, and then damaged during a collision at Finschhafen on 13 June 1944. It was yet again made airworthy, before finally being condemned and withdrawn from service at the end of the war.

Profile 45 – C-47 serial 42-23617, squadron number 156, *Bad (Mag)gie*

This C-47 was delivered to the 57th TCS on 28 June 1943 at Brisbane, and it narrowly escaped disaster when it had a mid-air collision near Nadzab on 29 April 1944. It was repaired and returned to service, seeing out its final days based at Elmendorf Air Force Base, Alaska, from 1951 to 1954.

C-47 serial 42-23617 Bad (Mag)gie, the subject of Profile 45, at Nadzab.

Kings Cross Shuttle, possibly the most accident prone of all Fifth Air Force aircraft, being refueled at Sydney during a liaison visit there.

A US recovery team at the crash site of Shakes All Over in 1979 where the numbers "16" from "X163" can be seen. On the left is Papua New Guinea citizen Richard Leahy who led the team to the site. (courtesy Richard Leahy)

C-47A serial number 42-24410 Coral Princess was assigned to the 58th TCS on 13 November 1943. It survived the war.

CHAPTER 14
58th Troop Carrier Squadron

The 58th TCS was activated on 18 November 1942 at Bowman Field, Kentucky, as one of four squadrons making up the 375th TCG alongside the 55th, 56th and 57th TCS. The squadron's commander at inauguration was Captain Milton Hardeman, followed by Lieutenant Charles King from 5 May 1943 who served for over a year until replaced by Captain Thomas White on 14 May 1944.

After training in the US for overseas deployment which included practising air-drops and towing gliders, on 2 June 1943 the 58th TCS moved to Baer Field, Indiana, where it received combat equipment in preparation for the move to the SWPA. On 7 June its first C-47s arrived at Brisbane and after more training these moved forward to Port Moresby on 10 July. It then moved to Dobodura on 19 August but returned to Port Moresby on 21 December. It operated from Nadzab from 22 April 1944 before leaving the SWPA when it moved to Biak on 25 September 1944.

The 58th TCS lost five C-47s during its New Guinea operations, the last of which occurred on 17 November 1944 at Kainantu when squadron number 193 was written off in a landing accident.

C-47 Fat Cat, the subject of Profile 48, parked at an unidentified Queensland airfield, whilst Miss America (from the 41st TCS, see Profile 35) departs in the background.

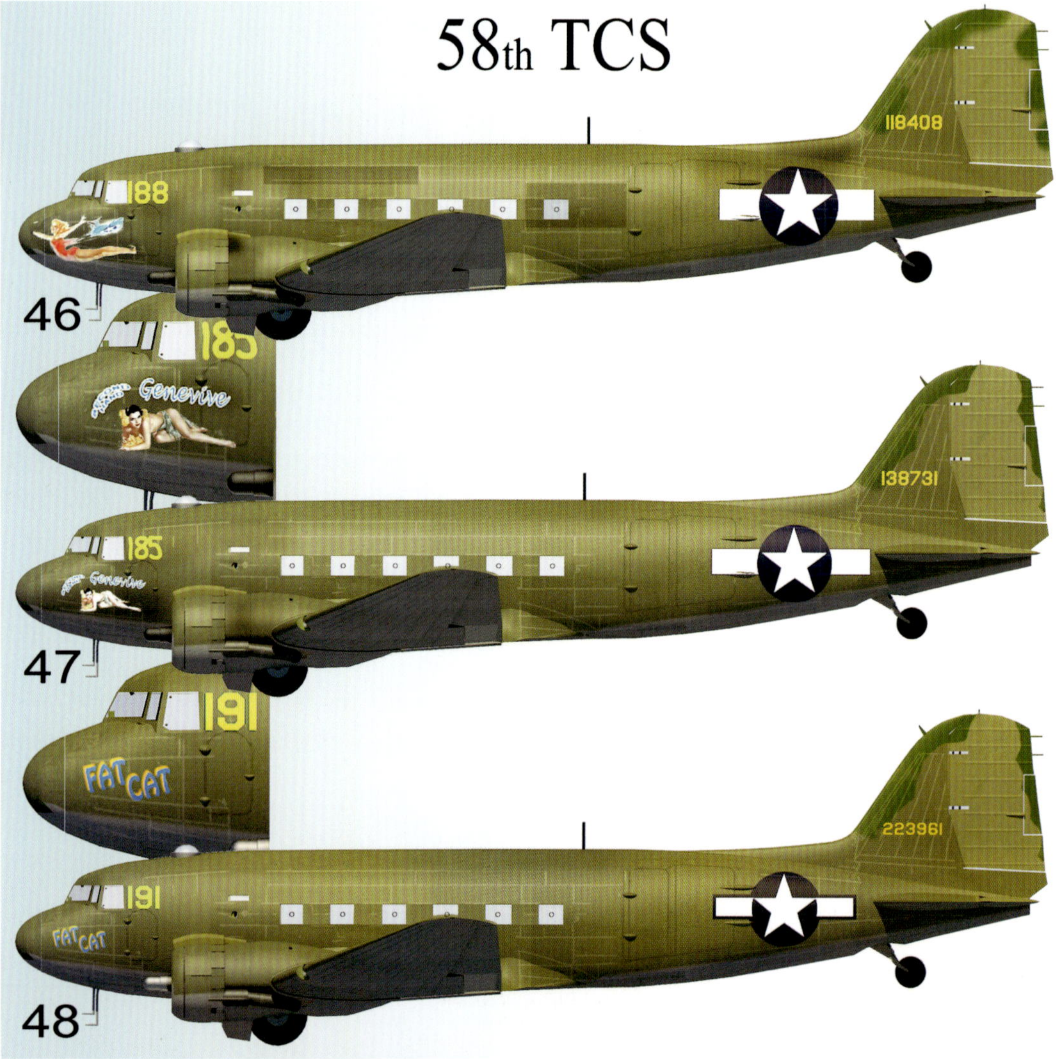

58th TCS

The nose art of C-47 serial number 41-18408, the subject of Profile 46, was a copy of the most popular Alberto Vargas pin-up girl of the Second World War. As soon as it had appeared in the 1943 Esquire Calendar this image instantly started appearing on USAAF aircraft world-wide. It still appears today on the occasional airliner.

Profile 46 – C-47 serial #41-18408, squadron number 188

This transport was a 58th TCS original which arrived at Brisbane on 13 July 1943 after crossing the Pacific. It suffered two repairable accidents during its New Guinea tour before returning to the US in October 1945. There it flew with civilian operators post-war before being sold to the Chilean Air Force which lost it in a crash on 12 May 1967.

Profile 47 – C-47 serial 41-38731, squadron number 185, *Second Hand Genevive*

This transport was another 58th TCS original which arrived at Brisbane on 11 July 1943 after crossing the Pacific. It was assigned to the Pacific Wing of Air Transport Command on 11 September 1944, before transferring to civilian service in the US in October 1945. It was still flying with airline Aérocargo Régional in Chile as late as 1997.

Profile 48 – C-47 serial 42-23961, squadron number 191, *Fat Cat*

Delivered to the 58th TCS on 14 August 1943, this transport was operated concurrently with the DAT callsign VHCHM. It was destroyed by fire on 6 November 1943 when strafed at Gusap by Japanese Ki-61 Tony fighters from the 78th *Sentai*.

Australian troops prepare to load aboard C-47 serial 41-38731 Second Hand Genevive, the subject of Profile 47, at Seven-Mile 'drome in late 1943. Note the white bars added to the original circular US insignia.

Both the subjects of Profiles 49 and 50 (serial numbers 43-15243 and 42-100456) are seen parked at Torokina with their cargo doors open in December 1943.

C-47 Peggy (serial number unknown) seen taxiing at Torokina.

CHAPTER 15
63rd Troop Carrier Squadron

Activated on 12 December 1942 at Bowman Field, Kentucky, the 63rd TCS was part of the 403rd TCG and immediately commenced training for overseas deployment with its C-47s. The first of these arrived at Espiritu Santo on 15 July 1943 where the squadron spent its first several months in theatre mainly flying support routes for the Thirteenth Air Depot. Its logo was a smirking *Donald Duck* in uniform watching supplies drop from a transport with a lightning bolt in the background.

Transport logistics during the Guadalcanal campaign had been reformed in late November 1942 before the 63rd TCS arrived in theatre. This was by the creation of South Pacific Combat Air Transport Command (SCAT), which was an autonomous command structure needed for the forecast increased support for upcoming Solomons operations following the end of the Guadalcanal campaign. SCAT was a conglomerate of both USMC and USAAF units, initially comprising VMJ-253 and its Headquarters Squadron, detachments from MAG-25 and the 13th TCS (which was later incorporated into the 403rd TCG). By February 1943 SCAT also included VMJ-152, VMJ-153, SMS-25 and the USAAF 801st Medical Air Evacuation Squadron.

Following its initial logistics role, the 63rd TCS was assigned to SCAT on 10 October 1943 in preparation for the Bougainville campaign. During this upcoming period, it would transport personnel, munitions, rations, spare parts and medical supplies. It would also conduct sundry liaison flights including the carriage of VIPs. Returning flights from front-line airfields carried wounded personnel, aircrew on leave for Australia or New Zealand, administrative minutiae and parts destined for overhaul including engines. Medical flights typically carried a nurse, a corpsman and sometimes a flight surgeon as part of the crew. The 63rd TCS remained with SCAT until 2 July 1944, after which its moved headquarters to Los Negros on 30 August 1944, and then to Biak in October.

The 63rd TCS incurred two major operational losses during its time in the SOPAC theatre, the worst of which was serial number 41-18675 lost on 24 November 1943. The aircraft (flown by a 64th TCS crew) departed from Tontouta at 0744 bound for Espiritu Santo. Captained by Lieutenant Philip Anders, the aircraft never arrived at its destination. Two days later pieces of its wreckage were collected from within Nakety Bay on the northern coast of New Caledonia about 50 miles northwest of Noumea. Most of the 25 onboard lost were Marines, in addition to three RNZAF personnel and the four USAAF crewmen. The second loss occurred on 4 February 1944 when C-47 serial number 42-24418 flown by Lieutenant Merrill Fink also went missing. With four crew and one passenger onboard, it was headed that morning for Pekoa on Espirito Santo from Tontouta in low cloud and persistent rain.

During its time in the SOPAC theatre the 63rd TCS did not apply squadron numbers.

63rd TCS

49

50

Profile 49 – C-47 serial #43-15243, *Ruby*

This transport was assigned into the 63rd TCS at Noumea on 11 April 1944. It was destroyed in an operational accident in the Philippines on 25 May 1945.

Profile 50 – C-47 serial #42-100456

Assigned into the 63rd TCS on 26 November 1943, this transport was photographed at Torokina on Bougainville shortly thereafter. The purpose of the single star on the tail is unknown. The light area on the leading edge of the fin is where the anti-icing boot has been removed.

CHAPTER 16
64th Troop Carrier Squadron

The history of the 64th TCS largely paralleled that of the 63rd TCS (see Chapter 15). Activated on 12 December 1942 at Bowman Field, Kentucky, the squadron was also part of the 403rd TCG. It immediately commenced training for overseas deployment with C-47s, the first of which arrived at Espiritu Santo in late July 1943. In contrast to the 63rd TCS, the 64th TCS focused on logistics support for combat units, mainly USMC Corsair squadrons operating from Turtle Bay airfield on Espirito Santo. The squadron logo was a stork dropping a baby holding a Tommy Gun.

The 64th TCS began contributing to SCAT missions in early August 1943, although it was not directly assigned to that command structure. Curiously, in 1964 the US Army Chief of Military History argued that the squadron never officially operated with SCAT, thus denying the squadron as an incumbent party to the Navy Unit Commendation awarded to SCAT. This countered its inclusion in the original citation.

Nonetheless all transport missions to forward areas directed by the USMC were coordinated by SCAT, and the 64th TCS often ferried SCAT personnel around the theatre. Furthermore, squadron crews were assigned to SCAT missions as required, especially during the peak demands of the New Georgia and Bougainville campaigns. Indeed, during the New Georgia campaign, 64th TCS ground crews prepared para-pack drops and were directly controlled by Commander Air Munda (COMAIRMUNDA).

The 64th TCS lost only three transports to operational accidents during its time in SOPAC, the most unusual of which occurred on 5 September 1943. The event produced a remarkable survival story after one of its C-47s departed Guadalcanal headed for Espirito Santo. Lieutenant Robert Healy was commanding a crew of five flying C-47 serial number 42-23711 which was a 64th TCS original delivered to the theatre by the same crew flying it that day. It was due to arrive at Turtle Bay airfield by 1630 but instead crashed into jungle near Mount Turi in bad weather, about fifteen miles northwest of its destination. The wreckage was located by villagers two days later.

Against all odds, one crewmember survived the crash, navigator Lieutenant Leonard Richardson. He had been sitting behind the cockpit bulkhead at time of impact and was badly wounded, suffering broken limbs and head injuries. It appears likely that the pilots let down in marginal visibility on what they assumed was a track for Turtle Bay. However, charts of the time were inaccurate, especially in respect to spot heights, and the C-47 drove cleanly into terrain at a descent speed of around 170 knots. There was no fire, largely as the aircraft was almost out of fuel. Following his recovery in hospital, Richardson walked back to the village where he presented the residents with a white mug as a token of gratitude for saving his life. The mug has been passed down through the generations where its custodianship is still proudly held today.

The second loss suffered by the 64th TCS was C-47 serial number 41-38742, destroyed during a take-off accident at Espirito Santo on 28 November 1943. The third loss occurred on 26

64th TCS

51

TEXAS TRAMP

52

Profile 51 – C-47 (unknown serial), *Mickey McGuire*

Profiled at is appeared in December 1943 at Espirito Santo, the serial number of this transport is unknown although it fell in either the 42-23XXX or 41-38XXX range. It was later named *Mickey McGuire* and had the unit logo painted on the starboard nose.

Profile 52 – C-47 serial #42-23722, *Texas Tramp*

Texas Tramp was a 64th TCS original which arrived in the South Pacific theatre on 23 July 1943. On 12 January 1944 it suffered an undercarriage collapse at Espirito Santo requiring a double engine change. It was lost a fortnight later on 26 January 1944 as described above.

C-47 serial number 42-23722 Texas Tramp, the subject of Profile 52, after its accident at Espirito Santo on 12 January 1944 which necessitated a double engine change.

January 1944 when C-47 serial number 42-23722 *Texas Tramp* departed Espiritu Santo bound for Guadalcanal after a double engine change following a wheels-up landing. An oil line broke on one of the replaced engines and the pilot ditched the C-47 in a sheltered bay on the north-western tip of San Cristobal Island. No injuries were sustained, and forewarned by radio, all survivors were collected by PBY the following day.

Similar to its sister squadron, the 63rd TCS, the 64th TCS did not apply squadron numbers to its aircraft during its time in the South Pacific.

C-47 serial number 42-23722 Texas Tramp, the subject of Profile 52, following its ditching off San Cristobal on 26 January 1944.

The fresh nose-art of Profile 51, just before the name Mickey McGuire was applied.

Ground crew pose in front of C-47 serial #42-24228 Our Lillian Ethel / Form 1A, the subject of Profile 54, prior to its loss on 6 March 1944.

CHAPTER 17
65th Troop Carrier Squadron

Activated on 12 December 1942 at Bowman Field, Kentucky, the 65th TCS led by Lieutenant Malcolm Long was originally part of the 403rd TCG training for overseas deployment with C-47s. However, due to logistical needs in New Guinea the squadron was reassigned to the Fifth Air Force just after it reached Australia on 26 July 1943. There it operated with DAT, before assignment to the 433rd TCG on 9 November 1943. The squadron did not adopt a logo during its wartime service.

The 65th TCS established its headquarters at Port Moresby in late July 1943 and also established a detachment at Tsili Tsili on 18 September 1943, just after participating in the Nadzab air drop. It moved its headquarters to Nadzab on 9 October 1943 whilst maintaining the Tsili Tsili detachment until 31 October 1943. It later established a forward detachment at Tadji, on New Guinea's northern coast, from 3 May to 2 June 1944. It moved out of the SWPA to Biak on 18 October 1944, and thence the Philippines from 24 January 1945. Major Marvin Calliham took over as commander on 27 December 1942 and led the unit throughout its New Guinea era, not replaced until 13 November 1944 by Major Donald Anderson. Captain Vernon Guess remained the squadron executive officer during this time.

The 65th TCS lost three C-47s to operational causes during its time in the SWPA. On 8 December 1943 serial number 42-23851 ditched in the Huon Gulf, just off Lae, with all crew rescued (see Profile 53). The squadron lost its first crewmen on the afternoon of 19 February 1944 when Lieutenant Paul Bishop radioed that he was having engine trouble with serial number 42-24398 between Dobodura and Finschhafen. Two of the four crew were killed when the transport subsequently force-landed inland from Cape Ward Hunt. The location was fortunately pin-pointed by B-24 crew members in the vicinity whose navigator plotted the crash location. After several unsuccessful attempts to reach the site four paratroopers from the 503rd Parachute Infantry Regiment jumped close to the downed plane and administered medical aid. Two badly wounded crew were carried out by natives on stretchers.

On 6 March 1944 the third 65th TCS loss occurred when C-47 serial number 42-24228 went missing when flying from Finschhafen to Saidor with thirteen people onboard. The pilot, Lieutenant John Hutchinson, had radioed to others in his formation at 1845 that they should turn around due to impending darkness. Hutchinson had been flying at only 700 feet and subsequently flew into the foothills of the Finesterre Ranges near Orarako village. In 1949 a US team recovered the remains of the crew.

65th TCS

Profile 53 – C-47 serial #42-23851, squadron number 213

This C-47 was a 65th TCS original flown to Australia on 26 July 1943. It participated in the Nadzab air drop as part of A Flight but was later lost off Lae on 8 December 1943. Note the red surround on the US insignia.

Profile 54 – C-47 serial #42-24228, squadron number 201, *Our Lillian Ethel / Form 1A*

This transport was delivered to the 65th TCS at Brisbane on 24 September 1943 and was lost with four crew and nine passengers on 6 March 1944 as described above.

A USAAF C-47 undertakes medical evacuations from Bougainville in early 1944, with a nurse visible on the left who was part of the additional medical crew usually carried on such flights.

A C-47 departs Wards 'drome, Port Moresby, in early 1943. The airfield complex was then undergoing substantial expansion of its taxiways and revetments.

C-47A serial number 42-23720 Sandy, the subject of Profile 56, being refuelled at Lae.

CHAPTER 18
66ᵗʰ Troop Carrier Squadron

As with the 65ᵗʰ TCS, the 66ᵗʰ TCS was activated on 12 December 1942 at Bowman Field, Kentucky. Led by Major Donald Smith, the 66ᵗʰ TCS was originally part of the 403ʳᵈ TCG which received its first C-47 on 16 January 1943. The unit focused considerably on glider-tow training during its initial period, however due to the theatre demands of New Guinea, the squadron was reassigned to the Fifth Air Force just after it reached Australia on 21 July 1943. It flew its first mission on 28 July 1943, a supply drop to Australian troops near the village of Mubo. For its first few weeks of operations, it flew under the guidance of 21ˢᵗ TCS crews who had already acquired experience in the field. The 66ᵗʰ TCS was assigned to the 433ʳᵈ TCG on 9 November 1943 and later operated with DAT. Its logo was a flying boxcar ridden by a smiling armed soldier.

The 66th TCS established its headquarters at Port Moresby in late July 1943 led by Smith's replacement Major Donald Woods and assisted by executive officer Captain James McIllivray. The squadron participated in the Nadzab air-drop after which it moved its headquarters there on 25 September 1943. It later established a forward detachment at Tadji, on New Guinea's northern coast, from 12 May to 2 June 1944, where it joined a 65ᵗʰ TCS detachment there. Its purpose at Tadji was to assist Fifth Air Force units in the move to Hollandia. The 66ᵗʰ TCS left the SWPA for Biak on 18 November 1944, and thence operated in the Philippines from 18 January 1945.

The 66ᵗʰ TCS lost a disproportionate number of C-47s, with ten lost to operational accidents in the SWPA. The most unusual operational loss unfolded on 10 August 1943 when serial number 42-23700 (squadron number 233) collided with a C-47 (squadron number 168) operated by the 21ˢᵗ TCS. The 66ᵗʰ TCS aircraft made a successful belly landing in a swamp about eight miles north of Terapo, while 168 returned safely to Seven-Mile. The downed crew were rescued several days later, and at time of publication the downed C-47 is still extant in the swamp.

An encounter with enemy fighters occurred on 9 November 1943. Lieutenant JD Larson was flying serial number 42-23723 (squadron number 239) from Nadzab to Dumpu when it was attacked by a pair of 59ᵗʰ *Sentai* Ki-43-II Oscar fighters, wounding two of the crew. The riddled transport then force-landed in the Markham Valley some 22 miles from Nadzab. Most of the hits had been to the radio/navigator's compartment, and the aircraft was later flown out and repaired.

66th TCS

229
OKLAHOMA LIMITED
2100482
55

238
SANDY
223720
56

231
MAGGIE
223694
57

237
BILLIE
223713
58

Profile 55 – C-47A serial #42-100482, squadron number 229, *Oklahoma Limited*

This C-47A was assigned into the 66th TCS on 7 December 1943. It was lost on 7 August 1944 when its port engine failed *en route* from Wakde to Finschhafen, forcing it to land on Tami beach not far from Hollandia.

Profile 56 – C-47A serial #42-23720, squadron number 238, *Sandy*

Assigned as a brand-new aircraft to the 66th TCS in the US on 16 July 1943, this transport was then flown to Australia. It participated in the Nadzab operation as one of A Flight. It was subsequently transferred to the 58th TCS, however on 14 August 1944 it collided with an A-20G on final approach into Nadzab from Finschhafen. The C-47 was force-landed at Nadzab without crew injuries but was written off.

Profile 57 – C-47A serial #42-23694, squadron number 231, *Maggie*

This was another transport assigned as a brand-new aircraft into the 66th TCS in the US on 16 July 1943, before being flown to Australia. It participated in the Nadzab operation as part of A Flight. On 26 October 1943 it suffered a landing accident at Dumpu, tearing the port engine from its mounts, and the aircraft was written off.

Profile 58 – C-47A serial #42-23713, squadron number 237, *Billie L*

This C-47A was assigned as a brand-new aircraft into the 66th TCS in the US on 16 July 1943. It was then flown to Australia where it participated in the Nadzab operation as part of A Flight. On 12 February 1944 it was running supplies to Kerowagi from Nadzab with four crew and five passengers onboard. On final approach, the pilot, Lieutenant Paul Jacobs, realised the airfield was not Kerowagi, and tried to go around. Despite applying full power, the aircraft settled on the runway at Mingendi Mission at about the halfway point. It then struck a building with its wing and lurched over an embankment, breaking the fuselage as it caught fire. Although there were only slight crew injuries, the aircraft was a write off. Its replacement was named *Ghost of Billie L* to honour of the original. The aircraft is profiled as it appeared at Finschhafen in late 1943. Note the mission and parachute marker(s), the latter representing its participation in the Nadzab drop.

C-47 squadron number 227 Poppy is seen from the shade of a refreshment shack at Kerowagi.

C-47A serial number 42-23713 Billie L, the subject of Profile 58, seen at Finschhafen with open cargo doors prior to its loss in February 1944.

The nose art on the aircraft which replaced Billie L. Its serial number is unknown.

C-47A serial number 42-23694 Maggie, the subject of Profile 57, after its accident at Dumpu. At the time of publication parts of the fuselage were still extant.

C-47 serial #42-24399, the subject of Profile 59, at Nadzab.

C-47 serial #42-23660 Stud Duck, the subject of Profile 60, being refuelled at Nadzab.

CHAPTER 19
67th Troop Carrier Squadron

The 67th TCS was activated on 9 February 1943 at Florence, South Carolina, and assigned to the 433rd TCG, along with the 68th, 69th and 70th TCS. Its C-47s moved to Baer Field, Indiana, on 1 August 1943, where long-range tanks were fitted for the Pacific crossing. The first of its C-47s departed for Hamilton Field on 12 August led by Major Erwin Baird, followed by the other three squadrons. The first of these arrived at Brisbane several days later where all of the long-range tanks were removed before the C-47s proceeded to Port Moresby, with the first arriving on 27 August. Baird was replaced by Captain William Head, a former school principal, as commander just before the squadron made the Pacific crossing. The 67th TCS logo was a phoenix arising from red flames against a blue sky.

The 67th TCS deployed as soon as it arrived in New Guinea, tasked with a two-day support mission to Australian troops fighting near Kaiapit in early September 1943. It then delivered supplies throughout the rest of the month to help build the new airfield at Tsili Tsili. The squadron moved to Nadzab on 5 November 1943, and then to Hollandia on 10 July 1944. All fuel supplies were arriving via the wharf at recently captured Lae and the 67th TCS started operating non-stop rosters during daylight hours throughout November 1943 delivering fuel drums from Lae to Nadzab. This service ceased when a road was built to Nadzab from Lae in December 1943. The squadron was the first to land transports at Tadji in April 1944.

The 67th TCS lost nine C-47s to operational accidents in the SWPA, the most unusual loss being C-47A 42-92051 (squadron number 304). On 10 February 1944 Lieutenant Andrew Dobrisky was flying in a formation when he collided with a C-47 from the 65th TCS, bending the fin and rudder of his aircraft to the right. He successfully ditched off Cape Ward Hunt and although the aircraft floated for a good while before sinking, Dobrisky was killed during the water impact and went down with the wreck.

C-47 serial #42-23886 Veda, the subject of Profile 61, at Finschhafen.

67th TCS

224399

301

STUD DUCK
306

223660

STUD DUCK
306

305
VEDA

223886

305
VEDA

318
Yingle-Yangle

223653

318
Yingle-Yangle

59

60

61

62

Profile 59 – C-47 serial #42-24399, squadron number 301, (Dumbo cartoon)

This C-47 was a 67th TCS original which crashed into hills not far from Kaiapit between Lae and Gusap on 2 January 1944, with the loss of four crew and five passengers.

Profile 60 – C-47 serial #42-23660, squadron number 306, *Stud Duck*

This C-47 was first assigned to the 46th TCS in June 1943 where it was named *Becky* and allocated squadron number 83. It was later reassigned to the 67th TCS where it was allocated squadron number 306 and renamed *Stud Duck*. The airframe was written off at Cape Gloucester on 14 March 1944 when it blew a tyre on take-off.

Profile 61 – C-47 serial #42-23886, squadron number 305, *Veda*

This C-47 was a 67th TCS original which landed at Brisbane on 24 August 1943 following the Pacific crossing. It was damaged by strafing Japanese Army Air Force fighters at Gusap on 6 November 1943, and then had a taxiing accident at Hollandia on 18 June 1944. Following repairs, it was finally destroyed when the pilot lost control on take-off at Finschhafen on 26 September 1944, but there were no crew injuries.

Profile 62 – C-47 serial #42-23653, squadron number 318, *Yingle Yangle*

This C-47 was reassigned to the 67th TCS from the 41st TCS at Brisbane as squadron number 87. It was subsequently named *Yingle Yangle* and given the new squadron number 318. It survived the war.

C-47 serial #42-23653 Yingle Yangle, the subject of Profile 62, at Wards 'drome, Port Moresby.

C-47A serial number 42-92050, known as Flying Irish to its crew and the subject of Profile 64, is seen on Marston matting at Finschhafen.

CHAPTER 20
68ᵗʰ Troop Carrier Squadron

The 68ᵗʰ TCS was activated on 9 February 1943 at Florence, South Carolina, under the command of Lieutenant Richard Adams. It was assigned to the 433ʳᵈ TCG, along with the 67ᵗʰ, 69ᵗʰ and 70ᵗʰ TCS. Its C-47s moved to Baer Field, Indiana, on 1 August 1943 where long-range tanks were fitted for the Pacific crossing. Subsequently the squadron departed from Hamilton Field, California, for the long trip to Australia, led by Major Joseph Bonner. The first C-47s arrived at Brisbane in late August before they proceeded to Port Moresby, arriving on 1 September. Three of the 68ᵗʰ TCS's original C-47s survived long-term Pacific war service: *Steppin' Out* (originally *My Gal Val*), *Double Cross* and *Mabel*. The squadron used no known logo during the war.

The 68ᵗʰ TCS moved to Nadzab on 15 November 1943 and sent a forward detachment to operate from Tadji from 18 May to 4 June 1944. The squadron moved from the SWPA to Biak on 15 November 1944, however a small detachment remained behind at Nadzab where it operated from 15 November 1944 until 5 January 1945.

C-47A serial number 42-23863 Whooo! The Gray Ghost, the subject of Profile 63, at Finschhafen.

68th TCS

Profile 63 – C-47A serial #42-23863, squadron number 331, *Whooo! The Gray Ghost*

This was an original 68th TCS aircraft which arrived in Australia on 26 August 1943. It blew a tyre at Finschhafen on 21 March 1944 and was subsequently written off. The yellow nose was a decorative feature only, applied later during its SWPA service.

Profile 64 – C-47A serial #42-92050, squadron number 339, (*Flying Irish*)

This was an original 68th TCS aircraft which arrived in Australia on 26 August 1943. It was destroyed when it caught fire on 27 May 1944. Although the aircraft never bore a name it was known affectionately by its crew as *Flying Irish*. It was mostly flown by Lieutenant Robert Stenglein who described his aircraft thus:

> … the rest of us succumbed to the crew chief's blarney and let Sergeant Ed Berry name our C-47 *Flying Irish*. Berry painted a handsome green shamrock on the nose. The Army identification gurus also insisted on painting large, yellow numbers on each side of the plane immediately behind the cockpit. Ours was 339.

The nose art on the port side of C-47A serial number 42-23860 Sure Skin, as seen at Finschhafen.

A fine study of an airborne Sure Skin just after departure from Finschhafen. Note that the red surround of the US insignia has been painted over with a dark colour, and the starboard side nose has the name but not the artwork carried on the port side.

C-47A serial number 42-23872 was named Defenceless Virgin and wore 69th TCS squadron number 356. It arrived at Brisbane from the US on 29 August 1943.

CHAPTER 21
69th Troop Carrier Squadron

The 69th TCS was activated on 9 February 1943 at Florence, South Carolina, under the initial command of Major Robert Flanigan. It was assigned to the 433rd TCG, along with the 67th, 68th and 70th TCS. Its C-47s moved to Baer Field, Indiana, on 1 August 1943, where long-range tanks were fitted for the Pacific crossing. The first of its C-47s departed for Hamilton Field, California, in mid-August in a move closely coordinated with the other three squadrons. The first C-47s subsequently arrived at Brisbane several days later before they proceeded to Port Moresby, arriving in stages around mid-September 1943. Flanigan divided the 69th TCS administratively into three flights: A, B and C. The squadron logo was a flying turtle carrying a package and 69th TCS aircraft were allocated squadron numbers in the range 351 to 375.

The 69th TCS was unique insofar as it was the only transport squadron in the SWPA not to incur fatalities. It came close on 15 January 1944 when Lieutenant Mayhew Fishburn was attacked by 59th *Sentai* Ki-43-IIs not far from Nadzab. Fishburn managed to execute an emergency landing, in the process becoming the only crew member wounded from the attack. For a while in 1944 the 69th TCS also operated B-17Es as transport aircraft, including a former 43rd BG aircraft named *Cap'n and the Kids.*

The 69th TCS moved to Nadzab on 15 November 1943 and sent a forward detachment to operate from Tadji from 18 May to 4 June 1944. The squadron moved from the SWPA to Biak on 15 November 1944, however a small detachment remained behind at Nadzab where it operated from 15 November 1944 to 5 January 1945.

69th TCS

65

66

Profile 65 – C-47A serial #42-24219, squadron number 366

This 69th TCS original arrived at Brisbane on 23 September 1943. It survived the war only to be lost when it crashed into a tree on 29 May 1973 in Canada killing all four civilian crew onboard. Note the white trim tab, a marking applied in the theatre around late 1944, and the over-painted red surround to the US insignia.

Profile 66 – C-47A serial #42-24218, squadron number 365

This 69th TCS original arrived at Brisbane on 23 September 1943 (in company with the subject of Profile 65). It was destroyed while parked at Nadzab on 28 April 1944. It also had a white trim tab, and an over-painted red surround to the US insignia.

The 69th TCS flight line at Finschhafen in early 1944.

C-47A serial #42-23919, the subject of Profile 67, seen at Nadzab just before it was lost.

CHAPTER 22
70th Troop Carrier Squadron

The 70th TCS was activated on 9 February 1943 at Florence, South Carolina, under the initial command of Major John McDonald where it was assigned to the 433rd TCG, alongside the 67th, 68th and 69th TCS. The squadron logo was a winged airman striding between islands carrying a large package.

The 70th TCS C-47As moved to Baer Field, Indiana, on 1 August 1943, where long-range tanks were fitted for the Pacific crossing. The first of its C-47As departed for Hamilton Field, California, on 24 August in close coordination with the other three squadrons. The first C-47A arrived at Brisbane several days later before they proceeded to Port Moresby, arriving throughout early to mid-September 1943. Once at Port Moresby, McDonald divided the squadron administratively into three flights: 1, 2 and 3.

The 70th TCS moved to Nadzab #2 strip in mid-November 1943 where each C-47A commenced on average a dozen trips a day between Gusap, Nadzab and Lae. It set a squadron record for a one-day airlift during this period with 512,000 pounds carried. Few 70th TCS C-47s carried names or nose art. Whether this was by design (a purist commander perhaps?) or just circumstance is unknown.

The 70th TCS lost seven C-47As in the SWPA to operational causes. The biggest loss of life occurred when C-47A serial 42-92062 (squadron number 385) went missing on 1 October 1944 with 24 personnel onboard. Flown by Lieutenant Russell Morrison, it had departed Hollandia and was bound for Finschhafen.

70th TCS

67

68

Profile 67 – C-47A serial #42-23919, squadron number 379

This transport was an original 70th TCS aircraft flown to Australia on 24 August 1943. On 17 September 1943 it arrived in Cooktown's circuit area *en route* from Townsville to Port Moresby and found the airfield to be fogged in. It nonetheless landed at Cooktown where it ran off the runway into soft ground. The airframe was written off when the undercarriage collapsed.

Profile 68 – C-47A serial #42-100723, squadron number 388

On 6 March 1944 this transport became the 70th TCS's most unusual loss. It was ferrying USAAF personnel to Sydney from New Guinea after a stopover at Townsville for rest and recreation leave. Most of the passengers were fighter pilots and the aircraft hit trouble over Cessnock, New South Wales, at 14,000 feet when the starboard propeller broke off and imbedded itself in the wing. The plane rapidly lost height, and the pilot Lieutenant Thomas McGinnis ordered the passengers to bale out, and for most of them it would be their first jump. When this order came through the altitude had dropped to about 3,000 feet. Fifteen jumped, however the parachutes of the last two failed to deploy. With his starboard wing badly torn, the starboard engine vibrated severely on torn mounts. However, McGinnis succeeded in making a successful forced landing in a small clearing. Despite the starboard cowling being forced into the cockpit under the pilot's legs when the engine tore loose, the five remaining onboard escaped major injury.

C-47A serial #42-100723, the subject of Profile 68, following its crash on 6 March 1944 near Cessnock, New South Wales.

C-47 VHCFQ was 41-18597, seen here at Wards 'drome in December 1943.

C-50 41-7697 as operated by Guinea Airways undergoes maintenance at Garbutt in 1944.

CHAPTER 23
Directorate of Air Transport

Understanding both the creation and functions of the Directorate of Air Transport is central to explaining its markings and operations in the SWPA. Commonly known throughout the theatre by its acronym DAT, this organisation was created from Air Transport Command at Amberley near Brisbane on 28 January 1942. The change was needed in order to avoid confusion with the South Pacific Wing of Air Transport Command then servicing trans-Pacific routes, and with which DAT bore no relation. Although an administrative creation, DAT nonetheless adopted a unique markings system. Regardless of their assignment to DAT or otherwise, C-47 squadrons remained under the operational command of their parent unit.

DAT underwent several metamorphoses during its existence and prior to the creation of the Fifth Air Force on 3 September 1942, it operated both USAAF and RAAF aircraft. DAT was then placed under the overall command of Major-General George Brett, with Group Captain Harold Gatty appointed its first commanding officer. Gatty was a civilian specially commissioned into the RAAF in order to legitimise his military authority. DAT initially acquired a wide range of aircraft types, however, the majority were Douglas transports the first of which were C-39s and C-53s. The latter had originally been bound for the Philippines but were diverted to Brisbane after the start of the Pacific War.

Pilots from Class 41H had only been a week at sea *en route* to the Philippines from Hawaii when they heard of the attack against Pearl Harbor. None had multiengine time, yet within the next few months several were assigned to a group which would shortly be constituted as the 22nd Transport Squadron at Essendon Airport in Melbourne. These pilots quickly learned how vast Australia and the Southwest Pacific were. After four hundred hours as co-pilots, they were deemed ready to become pilots in command. In March 1942 the first C-53 flights commenced to Port Moresby, the principal cargo being 0.50-inch calibre ammunition for No. 75 Squadron's Kittyhawk fighters. These trips were timed to arrive at dusk and to depart before dawn to avoid being strafed by Tainan *Kokutai* Zeros based at Lae.

During early February 1942 these transports also assisted with the evacuation of military and civilian personnel from the Netherlands East Indies. While this evacuation was in process, DAT transports were also flying supplies and personnel to strategic points within Australia. The C-53s delivered the US Army 102nd Coast Artillery (Anti-Aircraft) Battalion to Darwin in twelve days. On return journeys from New Guinea to Australia the transports carried wounded soldiers.

Responding to a request from the Pentagon that a USAAF transport squadron be activated in Australia, on 3 April 1942 the 21st and 22nd Transport Squadrons were created. The latter was based at Essendon Airdrome and its personnel included veteran combat pilots from the Philippines alongside a cadre of enlisted men from the US. Several transports including DC-3s purchased from the Dutch airline KNILM were assigned to this unit. Until late May 1942, these two squadrons confined operations as far north as Port Moresby, but aircraft flown by pilots from the 21st TS soon pressed further into New Guinea's interior on vital resupply missions to Australian forces at Wau. On 22 May 1942 the 21st TS ferried troops and supplies into Wau and Bulolo, escorted by P-39 Airacobra fighters. One DAT C-49 was specifically assigned to the Red Cross for the evacuation of wounded, and aerial drops of medical supplies to front-line stations.

Against the threat of Japanese fighters, bad weather and hostile geography, DAT transports dropped supplies to forward jungle positions throughout the Papua campaign, a large percentage of which were not recovered due to impenetrable jungle. On the final drive to Buna, air transport played a prominent role in the Kokoda and Buna campaigns.

In August 1942 it was recommended that all air transport activities in the SWPA be placed under DAT authority, however such over-reaching control by definition included delivery prioritisation. This was a key consideration as there was increasing demand on the limited air transport facilities within Australia, including on the three Australian civilian airlines. The eventual agreement placed these civilian operations under military control, meaning these could be flown by both military crews and/or their civilian counterparts. These airliners also availed themselves of the widespread military support and service facilities throughout the theatre.

On 12 November 1942, the 21st and 22nd TCS were reassigned into the newly designated 374th TCG, alongside the 6th and 33rd TCS (both of which had recently arrived in the SWPA). The two latter squadrons completed moves to Port Moresby by 28 December 1942, whereas the 21st and 22nd TCS relocated to Wards 'drome in February 1943. During this time a plethora of different types continued to operate alongside the Douglas transports including C-56s and DC-5s.

Peculiarly, DAT control stations along the transport routes were staffed by US personnel whereas the loading of aircraft was the responsibility of Australian units. DAT control headquarters for New Guinea operations moved around Port Moresby's airfields too; on 15 July 1942, the first opened at Seven-Mile 'drome, then briefly relocated to Kila 'drome, then returned to Seven-Mile before relocating to Wards 'drome where it stayed until commencement of the Philippines campaign. On 26 January 1943 Colonel Ray Elsmore was appointed DAT's new director to initiate a new expansion phase. Elsmore had years of civilian flying experience and was also a qualified lawyer. The colonel and his staff fell under the administrative control of Headquarters, Fifth Air Force. By this stage DAT had depots at Port Moresby, Melbourne, Adelaide, Brisbane, Townsville, Cooktown, Cairns, Horn Island, Iron Range, Darwin, Alice Springs, Daly Waters, Charleville and Rockhampton, delivering on average a thousand tons of freight and personnel per month. However, aircraft maintenance remained challenging, and at various times many transports remained grounded awaiting overhauls and/or spares.

By April 1943 additional DAT stations had been added at Onslow, Bachelor, Fenton, Iron Range, Daly Waters, Alice Springs, Perth and Milne Bay, and also Merauke in Dutch New Guinea. When Washington activated the 54th Troop Carrier Wing on 13 May 1943 to incorporate the arrival of new USAAF troop carrier groups, the responsibilities between this wing and DAT were delineated by Fifth Air Force Headquarters: DAT would haul materiel within the confines of Australia to the forward bases of Port Moresby, Milne Bay, Goodenough Island, and the Trobriand & Woodlark Islands. The 54th TCW was assigned to Advance Echelon (ADVON), Fifth Air Force, and was tasked to transport materiel into front-line positions. These locations gradually moved forward with the Allied advances, soon to include Dobodura, then Lae and Finschhafen, before progressing to Nadzab, Hollandia and on to Biak.

When the 317th TCG arrived at Port Moresby in January 1943 it was immediately placed under the operational control of DAT. This brought about the first of many markings complications in the theatre: all of the group's 56 brand new C-47s it had flown across the Pacific already had allocated three-digit squadron numbers, and most were decorated with nose-art and/or a name. In early February 1943 the group was ordered to hand them all over to the 374th TCG for front-line service. In exchange, the 317th TCG crews received the 374th TCG's worn C-47s. Subsequently a suite of other widespread transfers and reshuffles prioritised newer airframes to combat duty.

Whilst a *mélange* of other aircraft types also filled DAT's inventory, by July 1943 the 317th TCG was operating the C-47 exclusively. From 1943 the RAAF transport squadrons too began receiving new C-47s, giving them a massive increase in capability. By September 1944, the month it was dissolved, DAT operations had reached its zenith. It was operating throughout the entire Australian mainland, most of New Guinea, The Admiralty Islands, Bougainville, Goodenough Island, and the Dutch New Guinea locations of Biak, Wakde and Noemfoor. At its peak DAT serviced an average of 140 aircraft movements daily, and in addition to the 317th TCG it included the 6th, 21st and 33rd TCS, plus Nos. 33, 34 and 35 Squadrons, RAAF. Completing the DAT inventory were various other transport types, Douglas and otherwise, operated by Australian National Airways, Qantas and Guinea Airways.

DAT Markings

Whilst an orderly squadron numbering system for the US transport squadrons was intended, things did not work out this way. Major inventory reshuffles such as that between the 317th and 374th TCGs as outlined above scrambled these as did recurrent inventory reshuffles between units including service squadrons. Thus a squadron number, normally painted behind the cockpit windows of an aircraft, does not necessarily predicate its unit assignment. Furthermore, some C-47s retained the three-digit numbers first allocated in the US by the 317th TCG. Lazy scholarship combined with this unique set of circumstances has disappointingly led to ubiquitous colour profiles over the years identified against wrong units, serial numbers, Australian call signs and even nose art.

DAT thus implemented an identification system for all transports regardless of which air force or USAAF unit they served, by implementing radio callsigns in the VHC(XX) series. Although the civilian prefix VH representing Australia was already extant, the basis to allocate aircraft under military control a civilian callsign remains unclear. The Australian Department of Civil Aviation assigned this range, stemming from its existing registration sequence which had reached VHACX by 1942. In order to differentiate DAT registrations from civilian ones, it was ordered that the hyphen be omitted after the VH prefix, however some units ignored this requirement. As these DAT callsigns were not civil registrations, no Certificates of Registration were issued for relevant aircraft, and the call signs were not recorded in the Australian Civil Aircraft Register in Melbourne. When the VHC(XX) series was almost fully allocated, the VHD(XX) series commenced and appeared on several RAAF C-47s before it ceased.

DAT

Profiles 69 and 70 – DC-3 PK-ALW / callsign VHCXE

Douglas DC-3 PK-ALW conducted the final evacuation flight from the NEI on 7 March 1942, when KNILM airline pilot Eddy Dunlop took off from a road near Bandoeng after midnight and flew to Port Hedland, Western Australia, after seven and a half hours in the air. The transport was assigned to the USAAF at Archerfield on 15 May 1942 with the temporary callsign "11944" and then reassigned to the 374th TCG as VHCXE. In July 1942 it was fitted out as a VIP aircraft with a more spacious cabin and seven passenger seats, and was briefly used as General MacArthur's aircraft in Melbourne.

Profile 71 – C-50 serial 41-7697, callsign VHCDK, Guinea Airways

This transport is profiled in its original Guinea Airways camouflage scheme as it appeared at its home base of Parafield, South Australia, in late 1943. Note the white bars added to the US insignia. It later acquired the names *Laka Nookie* and *Kqitchyr Bichin'*. In late 1944 it was stripped down to natural metal finish and given field number W7697 (the last four digits of the serial number) in a similar markings scheme to that of Profile 75.

Profile 72 – C-53 serial #41-20051, tail code 62TG54

At the start of the Pacific War, two DC-3s operating with Australian National Airways became the most critical transport aircraft in Australia, joined in January 1942 joined by five USAAC Douglas C-53s which had arrived in Australia by ship. This C-53 had tail marking 62TG54, indicating it previously served as aircraft #54 with the 62nd Troop Carrier Group in the US. From January to March 1942, the small C-53 fleet did the bulk of military transport work in Australia until a larger fleet of ex-Dutch transports arrived. This C-53 was also allocated callsign VHCDX but was destroyed by strafing Zeros from No. 3 *Kokutai* at Bathurst Island near Darwin in February 1942.

The top half of Puddle Jumper's nose art (serial number unknown) has been painted over to eradicate a previous squadron number for DAT service.

DAT

73

74

75

76

Profile 73 – C-47A serial #42-92802, callsign VHCJT, *Fair Dinkum*

This transport is profiled as it appeared at Vivigani, Goodenough Island, in late 1944, after it had been transferred into the 21st TCS on 9 April 1944. It suffered structural damage on 14 December 1944 due to severe turbulence and had to be grounded for extensive repairs. It was destroyed by strafing Japanese fighters on 27 December 1944. Note the white trim tab marking of the 21st TCS.

Profile 74 – C-47 serial #42-23583, callsign VHCGJ, squadron number 61

This C-47 was originally named *The Pathfinder* when received by the 41st TCS on 30 May 1943, but was later renamed *Flamingo* and is seen here when it served with DAT. The aircraft retained its 41st TCS squadron number 61, and later served Fifth Air Force Service Command at the end of the war.

Profile 75 – C-50 serial #41-7698, callsign VHCDJ (W7698), Guinea Airways

This C-50 carried the names *Winnie*, *Waltzing Matilda* and *Narth Nikka* during its career and served with the RAAF from June 1943 to April 1944, before being reassigned to Guinea Airways in May 1944. It is profiled as it appeared with Guinea Airways at Townsville in late 1944. Note the white bars added to the US insignia.

Profile 76 – C-47A serial #42-23588, callsign VHCGP, *Blonde Baby*

This C-47A was originally transferred into the 40th TCS as *Blonde Baby* on 8 June 1943 and then later operated with DAT. It was later transferred to the 39th TCS in January 1944 at Wards 'drome, and was finally destroyed in an accident in the Philippines at the end of the war.

Hollywood actor Gary Cooper flew aboard DAT C-47A VHCJH (serial number 42-92792) during a United Service Organization performance at Dobodura in late 1943.

DAT

77

78

79

80

Profile 77 – C-47 serial #42-23657, callsign VHCGT, *Looking Fer Trouble*

This transport was assigned to the 41st TCS in June 1943 as squadron number 63 where it was named *The Venice Short Line*. After participating in the Nadzab air drop on 5 September 1943 it was renamed *Lookin' Fer Trouble* when it served with DAT as VHCGT.

Profile 78 – C-47 serial #41-18577, callsign VHCFG, *Bugle Nose*

This transport first served with the 6th TCS in Australia from 10 October 1942. It was transferred to the 40th TCS on 17 January 1943 where it was operated by DAT as VHCCO, then as VHCFG (named *Yanks Delight*) and finally named *Bugle Nose* as seen here. It returned to the US in August 1944, but was lost during civilian service in Canada on 2 January 1951.

Profile 79 – C-47 serial #41-18564, callsign VHCCU, DAT number 2, *Flying Dutchman*

This transport was assigned into the 33rd TCS in Australia on 26 October 1942, retaining the squadron number 564 applied in the US. It was operating with DAT as VHCCU when it crashed into the side of Mt Obree in New Guinea on 10 November 1942 while delivering troops from Port Moresby to Pongani. A limited number of survivors walked out to get help, but some went missing. The aircraft remained missing until it was found 1967. In late 1942 DAT was using large numerals to assist loading at Port Moresby on the sides of C-47s for the airlift to Pongani, allocating 2 to this aircraft.

Profile 80 – C-47 serial #41-18588, callsign VHCFO, DAT number 11, *The Hiawatha*

Similar to Profile 79, this transport was assigned into the 33rd TCS in Australia on 26 October 1942 and was first operated by DAT as VHCCV, then later as VHCFO *The Hiawatha*. DAT allocated number 11 to this aircraft for the Pongani airlift in late 1942. After returning to the US in 1944 the aircraft wound up its days flying for civilian operators in North America.

A textbook example of DAT tail markings - serial number 41-18612 has been painted over and replaced by the stenciled callsign VHCFY.

C-50 callsign VHCXD at Seven-Mile in early 1942. The Airacobras in the background are from the 8th Fighter Group.

C-49 callsign VHCDE serial number 41-7693 appears in natural metal finish at Saidor in late 1944, as operated by Australian National Airways under DAT guidance.

VHCDM retained its serial number, in an unusual variation of DAT markings.

C-50 callsign VHCDK, the subject of Profile 71, at the eastern end of Seven-Mile 'drome in late 1944 after it was stripped down to natural finish and given field number W7697.

Evelyn Lee was in DAT service when this photo was taken at Garbutt, but its serial number is unknown.

Douglas C-47B serial number 43-16300, the subject of Profile 82, at Biak about to depart on a flight to Australia.

Douglas C-47A serial number 42-23419, the subject of Profile 84, at Torokina in mid-1944.

CHAPTER 24
Miscellaneous C-47s

A British transport squadron operated from Parafield airport in South Australia in 1945. This was No. 238 Squadron, RAF, which was a large unit from India, operating 30 C-47As. It remained there until the start of 1946. Other miscellaneous operators of C-47s detailed here include USAAF bombardment groups and headquarters units.

Newly arrived C-47As of No. 238 Squadron, RAF, at Parafield in July 1945.

This late model C-47B, Arkansas Traveler [sic], is seen in the Philippines in June 1945 with red spinners. Operated by the Far East Air Force Headquarters, the badge on the nose incorporates both the Fifth and Thirteenth Air Force logos.

Miscellaneous

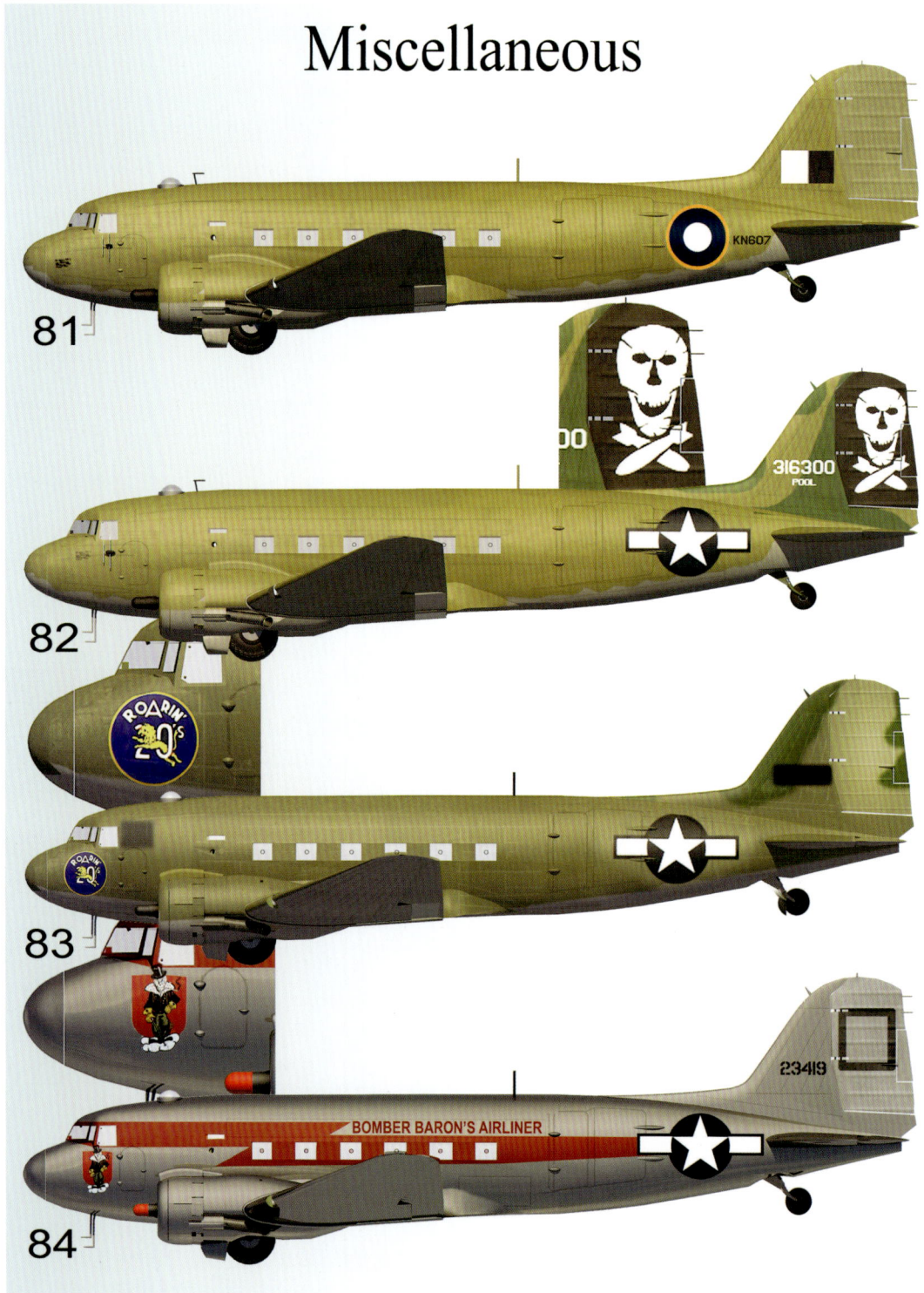

81

82

83

84

Profile 81 – Douglas C-47 Dakota KN607, No. 238 Squadron (RAF), Parafield, South Australia, July 1945

Towards the end of the war, Parafield airfield in South Australia received No. 238 Squadron, RAF, a transport unit from India operating 30 C-47As. The first three headed for South Australia on 26 June 1945 via Cumilla (Burma), Ratmalana (Ceylon), Cocos Island, Learmonth (Western Australia) and Guildford (Perth). The entire squadron had arrived at Parafield by 11 July, bringing 210 men while the ground echelon did not arrive by ship until 23 October. The squadron soon commenced regular flights between Australian and Allied bases in New Guinea and the Netherlands East Indies. The furthest location visited was Peleliu in the Central Pacific.

On 15 August (VJ Day) Parafield stood down for two days of celebration, and the end of hostilities presaged winding down No. 238 Squadron's presence in the theatre. Allied POWs started being repatriated to Australia at this time; in early October, some 66 ex-POWs were accommodated by the unit while *en route* to the eastern states. On Christmas Day it was learned that the squadron would be disbanded. Personnel were soon mobilised to embarkation depots for a return to the UK. During the first week of 1946, the squadron's C-47s commenced the long trip back to Britain while a detachment was diverted to Singapore.

Profile 82 – Douglas C-47B serial #43-16300, 90th Bombardment Group

This transport was assigned directly to the 90th Bombardment Group on 15 August 1944. The 90th BG was then operating B-24J Liberators and used the C-47B for supply and liaison flights throughout the theatre including visits to Australia. The distinctive skull and crossbones motif stencil on the rudder was a precise replica of that which appeared on the group's Liberators. This transport was lost to an accident in the Philippines on 6 March 1945.

Profile 83 – Douglas C-47B serial unknown, 312th Bombardment Group

This B-model C-47 was assigned to the 312th Bombardment Group in late 1944, then operating A-20Gs. The transport was used for supply and liaison flights throughout the SWPA, including to and from Australia. The *Roarin' 20s* lion motif on the nose was used by the group, reflecting the type of aircraft they flew in combat.

Profile 84 – Douglas C-47A serial #42-23419, 5th Bombardment Group

This transport was first assigned into the Fifth Air Force on 5 May 1943 but wound up serving the 5th BG of the Thirteenth Air Force as a hack aircraft. The unit's engineers stripped down the aircraft to natural metal finish and gave it airliner livery with the unit nickname *The Bomber Barons*. The last five digits of the serial number were stencilled on the fin. The aircraft flew throughout Indonesia after the war and was still flying as recently as 1989. Its current disposition is unclear. Note that in past other publications this aircraft has been incorrectly illustrated in blue livery instead of red.

A stretcher patient is loaded on an R4D at Barakoma in late 1943. Note the colour of the green zinc chromate etch primer interior.

CHAPTER 25
United States Marine Corps

The USMC transport units operated modified Douglas C-47s designated by the USN as the R4D series. The R4Ds incorporated a few differences to their USAAF counterparts. The R4D-1 had a reinforced metal floor with tie-downs and a 12-volt electrical system. The interior was configured with either folding wooden seats to carry 28 paratroopers, or with 18 slung stretchers and seats for three medical attendants. This latter configuration was common for SCAT medical evacuation flights.

The R4D series came equipped with a glider-towing cleat, a discernible feature which differentiated it from most C-47s. The upgraded R4D-5 model was rewired with a 24-volt electrical system and improved ducted cabin heating. All R4Ds had smaller carburettor intake air scoops than their C-47 counterparts, placed on top of the mid-cowl. A curiosity to arise from this era is that the nickname *Flying Boxcars* was widely used to denote SCAT R4Ds and C-47s, leading to the post-war designation of the Fairchild C-119 Flying Boxcar. A total of 23 R4Ds were lost to operational causes in both the SOPAC and SWPA theatres from 1942 to 1945, reflecting their widespread use.

This volume corrects the falsehood derived from lazy scholarship that R4Ds were allocated random two-digit numbers. They were hardly random, but instead derive from the last two digits of the BuAer serial number, accessible in the relevant unit records. These BuAer numbers used by USMC squadrons in the South Pacific fall in the following ranges: 01648 to 01984, 05052 to 05064, 12395 to 12436 and 37663 to 39090.

VMJ-253

Formed at Ewa, Hawaii, on 11 March 1942 under the command of Lieutenant Colonel Perry Smith, VMJ-253 was assigned to MAG-25 on 1 June 1942. It became the first USMC utility squadron to deploy to the SOPAC theatre, arriving in New Caledonia on 1 September 1942 and flying its first mission to Guadalcanal two days later. By February 1943 the squadron was operating a dozen R4D-1s.

VMJ-253 was thus one of the original SCAT Squadrons, with which it served until 15 June 1944. The squadron underwent numerous changes of command: Major Harold Johnson from 1 June to 11 October 1942, then Major Henry Lane until 14 October 1943, Major Freeman Williams until 3 November 1943, Major Douglas Keeler until 23 January 1944 and finally Major James Moran who was the last Commander to serve under SCAT.

VMJ-152

VMJ-152 was created in the US from VMJ-1 which had been disbanded on 7 July 1941. Its first three R4Ds departed San Diego on 10 October 1942 and arrived at Tontouta, New Caledonia, nine days later. Led to the theatre by Major Elmore Seeds, the squadron served SCAT for its entire existence until 20 July 1944, after which it was redesignated transport squadron VMR-152. Commanders who followed Seeds were Major Dwight Guillote from 14 February 1943, then Major Carl Fleps until 10 October 1943, Lieutenant Colonel (just promoted) Elmore Seeds until 14 January 1944, Lieutenant Colonel Frederick Leek until 1 June 1944 and then its final commander Lieutenant Colonel Albert Munsch until it was rebadged as VMR-152. Its complement in the SOPAC theatre was a dozen R4D-1s.

VMJ-153

VMJ-153 was established at San Diego on 1 March 1942 and was commanded by Major Ben Redfield. The squadron's R4Ds flew the Pacific route to New Caledonia in March 1943 followed by its support echelon in May by which time it was operating seven R4D-1s and three R4D-5s. The last USMC transport squadron assigned to SCAT; on 20 July 1944 the squadron was redesignated a transport squadron becoming VMR-153. Other commanding officers were Major Warren Sweetser from 6 August 1942 until 16 March 1943, then Major Harry Van Liew until 4 April 1943, Major William Lanman until 31 May 1943, Major Elmore Seeds until 4 July 1943, Major Robert Bell until 4 November 1943, Major Freeman Williams until 22 May 1944 and then its final commander under SCAT was Major Theodore Sanford.

SMS-25

Service Marine Squadron 25 was formed at Tontouta in New Caledonia under Major Leonard Ashwell on 18 November 1942, its purpose to provide maintenance and support to MAG-25 and SCAT. SMS-25 operated a handful of R4Ds throughout SOPAC outposts. Its other commanders were Captain Ralph Yeaman from 24 November 1942 to 15 November 1943, then Captain Jack Church until 9 March 1944.

MAG-25

Headquarters Squadron (HEDRON) of MAG-25, commanded by Captain Leroy James, operated a solitary R4D-5 (Bu 17159) throughout 1944.

VMJ-253 R4Ds in formation over the Solomons, accompanied by an SBD. The foremost R4D is squadron number 8.

This was one of the first SCAT R4Ds to land at Munda, squadron number 42.

USMC Squadrons

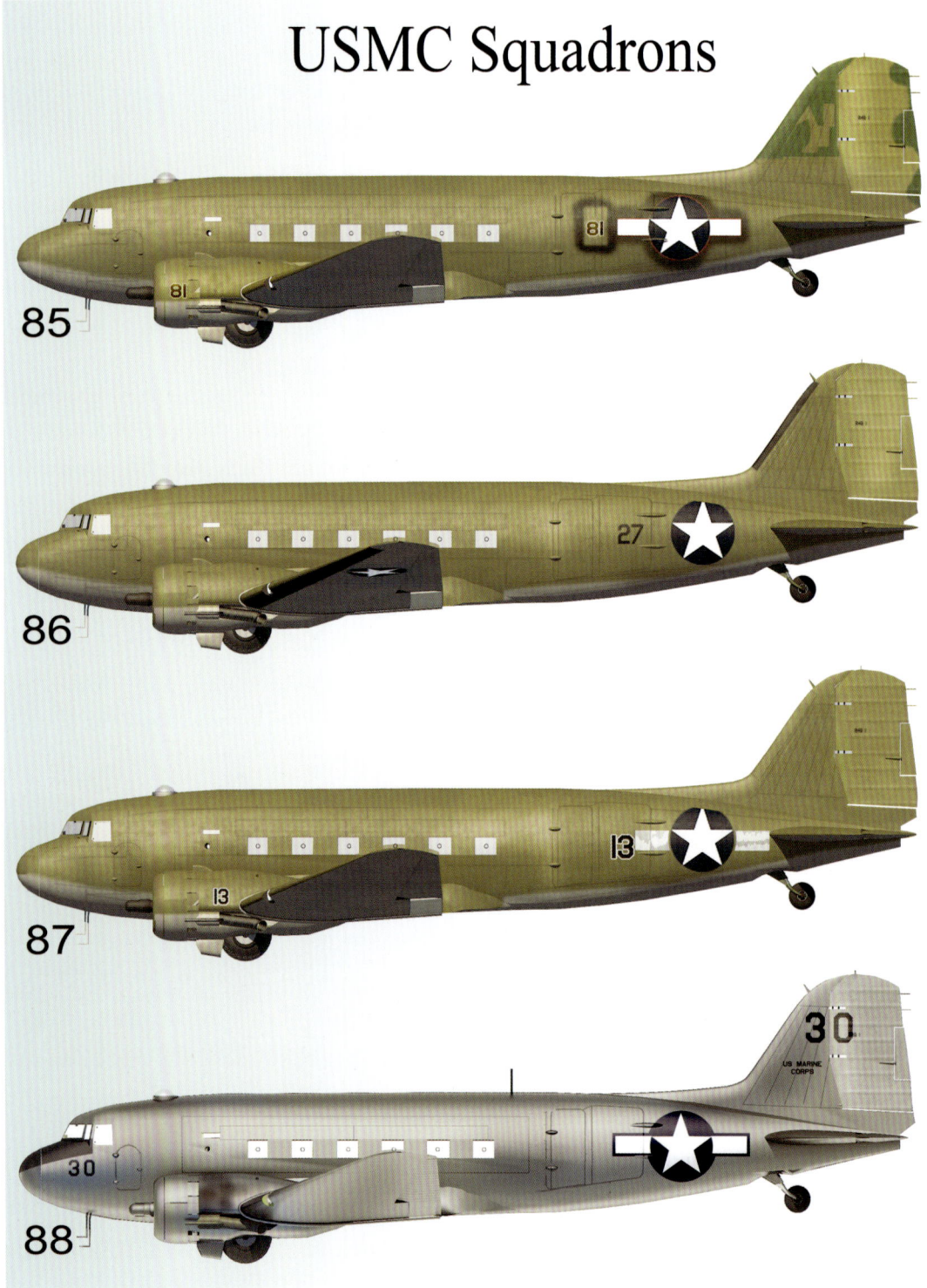

85

86

87

88

Profile 85 – R4D BuAer #01981, squadron number 81, VMJ-253

This R4D is profiled as it appeared at Guadalcanal in October 1942. The red outline of the US insignia has been painted over.

Profile 86 – R4D (BuAer unknown), squadron number 27, VMJ-152

This transport served with SCAT throughout 1942 to 1944.

Profile 87 – R4D BuAer #12413, squadron number 13, VMJ-253

This R4D is profiled as it appeared with SCAT on Guadalcanal in 1943, with white piping surrounding the number 13.

Profile 88 – R4D (BuAer unknown), squadron number 30, 1ˢᵗ Marine Air Wing

This transport is profiled as it appeared with the 1ˢᵗ Marine Air Wing on Bougainville in February 1945. Note the black anti-glare section in front of the windscreen.

An R4D of the 1ˢᵗ Marine Air Wing, number 30 and the subject of Profile 88, at Torokina in 1945.

USMC Squadrons

Profile 89 – R4D BuAer #12432, squadron number 32, VMJ-153

This R4D is profiled as it appeared in the Solomons in 1943. Note that the squadron number has been painted in white.

Profile 90 – R4D (BuAer unknown), squadron number 85, VMJ-253

This R4D is profiled as it appeared with SCAT on Guadalcanal in 1943. The dark areas mark painted-over repairs from combat damage.

R4D number 84 from VMJ-253 taxies on Marston matting at Torokina.

A USMC R4D heads out for Espirito Santo from Guadalcanal in early 1943.

C-47B A65-114, the subject of Profile 93, at an unknown Australian airfield.

The squadron code RE on the fuselage of C-47A A65-48 indicates assignment to No. 36 Squadron. It is conducting a low-level drop of supplies in New Guinea.

CHAPTER 26
Royal Australian Air Force

RAAF transport operations in the Pacific were extensive but mainly confined to the SWPA, with the occasional foray into Bougainville. The RAAF's involvement with Douglas transports commenced with the allocation of ten DC-2s in late 1940, several of which were former Eastern Airlines aircraft. Nos. 33, 34, 35 and 36 Squadrons were the first four RAAF dedicated transport squadrons, created in February/March 1942. Whereas No. 36 mainly operated DC-2s, the other three deployed mixed fleets of impressed civil aircraft and Short Empire flying boats. In July 1942 No. 36 Squadron came under the control of DAT. Its first major loss was DC-2 A30-5 which was assigned to the squadron on 14 September 1942 but was lost on its maiden flight that same night when it crashed on final approach to Seven-Mile killing all five aboard.

The delivery of two dozen C-47s in early 1943 saw the DC-2 fleet largely sidelined to training units. The first C-47s operated by the RAAF were loaned to No. 36 Squadron by the USAAF which from December 1942 was based at Townsville. Eight C-49s were also delivered to the RAAF in the next few months but these were reassigned to Australian airlines in May 1944 and replaced by thirteen C-47s direct from the US production line. The major influx of new C-47s (with A65 prefixes) were delivered to the RAAF from February 1943, and No. 36 Squadron used them to commence operations to New Guinea from Townsville in early June 1943. It soon established a detachment at Port Moresby, although initially most operations were limited to the Port Moresby-Townsville route.

No. 33 Squadron was formed at Townsville on 16 February 1942, with an inventory of Short Empire flying boats. It later operated from Port Moresby with an eclectic fleet including Ansons, De Havilland Dragons and Tiger Moths. It received its first C-47s in 1943 and operated them in New Guinea until the end of the war whereupon the squadron returned to Townsville where it was disbanded in May 1946.

No. 34 Squadron was formed at Darwin on 23 February 1942 with an initial inventory of impressed civilian transport aircraft. The squadron was reformed at Parafield in South Australia equipped mainly with Dragons and in May 1943 received its first C-47s. These flew between Parafield, Archerfield and Mascot, with later flights extending to locations such as Port Moresby, Finschhafen, Nadzab and Hollandia. In 1945 No. 34 Squadron moved to Morotai from where it became involved in the invasion of Borneo in May/June 1945.

No. 35 Squadron was formed at Pearce in Western Australia in March 1942 and operated in that area for most of the war. It received its first C-47s in August 1943 after which flights began ranging Australia-wide. Later regular courier services extended to New Guinea and Morotai. In March 1945 the squadron moved to Townsville.

No. 37 Squadron was formed at Laverton on 15 July 1943 and operated mainly Lockheed Lodestars throughout the war. It became the last RAAF squadron to receive the C-47, receiving its first in

February 1945 when it was based at Essendon, with two detachments at Parafield and Morotai.

No. 38 Squadron was formed on 15 September 1943 at Richmond, operating Lockheed Hudsons for the first eight months of its existence, before taking on C-47s in May 1944. It mainly flew internal Australian routes, before later flying to Hollandia, Biak and Noemfoor. In December 1944 it moved to Archerfield from where it continued flights throughout the SWPA. In July 1945 it sent a detachment to Morotai. It suffered only one fatal C-47 accident, a flight which went missing in Dutch New Guinea in September 1945 *en route* from Biak to Merauke. The squadron evacuated Australian POWs after the cessation of hostilities and operated the first RAAF aircraft to land at Singapore following the Japanese surrender.

When DAT was disbanded in late 1944 the RAAF resumed control of all Australian transport operations. In October 1944, in a major airlift, Nos. 34, 35 and 36 Squadron C-47s moved Beaufort and Kittyhawk squadrons from Higgins Field (near the tip of Cape York Peninsula) and Townsville to Tadji on New Guinea's northern coast.

Two No. 36 Squadron C-47s were stationed at Tadji from January 1945 to perform supply drops to Australian soldiers conducting mopping-up operations along the coast around Wewak. C-47 VHCUF crashed during one of these drops near But airfield, killing four crew. Four others were injured but were rescued by Australian commandos. A similar detachment of No. 36 Squadron C-47s was also stationed at Bougainville in early 1945 to perform air drops, while the squadron flew delivery flights as far as Morotai. By the end of the war, when USAAF transport units had left the SWPA, RAAF C-47 operations intensified from Townsville, and by early 1946 C-47 detachments were still operating from Darwin, Ambon and Morotai. A tri-weekly service to Rabaul via Port Moresby, Lae and Finschhafen was initiated by No. 38 Squadron C-47s in March 1945.

Overall the RAAF received 124 C-47s, C-47As and C-47Bs during war.

Markings

All RAAF C-47s were delivered in the standard USAAF Olive Drab/grey scheme but this often changed when airframes were rotated through workshops for overhauls and repairs. There was little standardisation of DAT and squadron markings, with *ad hoc* creativity the rule rather than the exception. Squadron codes were not always applied but if so, were designated as per this table:

Squadron	Code
33	BT
34	FD
35	BK
36	RE
37	OM
38	PK

C-47A A65-31 served with No. 36 Squadron using the DAT callsign VHCUC. Note the straight-line delineation between the Olive Drab and grey under surface.

C-47A A65-38, the subject of Profile 94, somewhere in New Guinea. Note the "H" behind the cockpit, being the last letter of its DAT callsign VHCUH.

RAAF

91

92

93

94

Profile 91 – C-47A serial #42-24139, A65-19, callsign VH-CTS, No. 35 Squadron

This C-47A was received on 23 August 1943 and assigned to No. 35 Squadron a month later where it was given DAT callsign VHCTS. It later flew with No. 33 Squadron and on 13 May 1945 after departing Lae for Hollandia it made an emergency landing at Madang when the starboard engine failed due to a faulty hose connection. After a post-war career with Australian airlines it was donated to the Cambodian Air Force in September 1971 but destroyed by the Khmer Rouge when they captured Phnom Penh in 1975.

Profile 92 – C-47A serial #42-92710, A65-42, L FD, callsign VHC-UL, No. 34 Squadron

This transport was issued DAT Callsign VHCUL as soon as it was delivered to the RAAF on 4 April 1944. It was mainly operated by No. 34 Squadron for the rest of the war. It served Australian and other regional airlines before being decommissioned in 1981.

Profile 93 – C-47B serial #44-77128, A65-114, RE E, callsign VHRGB, No. 36 Squadron

This C-47B was delivered to No. 36 Squadron late in the war on 23 June 1945 and given the DAT callsign VHRGB. It was repainted in Forest Green shortly after receipt. It crashed following a double engine failure on take-off from Edinburgh in October 1986 while still in RAAF service. It is currently on display at the South Australian Aviation Museum in Port Adelaide.

Profile 94 – C-47A serial #42-92447, A65-38, callsign VHCUH, No. 33 Squadron

Delivered to No. 33 Squadron on 18 February 1944 this C-47A carried the last letter of its DAT callsign behind the cockpit window. It served with No. 33 Squadron for most of its career and was allocated the fuselage code BT K. The aircraft was written off and abandoned in a take-off accident on 12 September 1945 at Hayfield strip in the Sepik district of New Guinea. At time of publication its fuselage still remains in kunai grass at Hayfield.

C-47A A65-42, the subject of Profile 92, in Australia's Northern Territory.

NZ3501, the subject of Profile 95, just after it was delivered to Whenupai in April 1943 with an unorthodox fuselage roundel applied over the US insignia.

NZ3505 served both Nos. 40 and 41 Squadrons and carried the nose art of a stork dropping a bundle with the name Anytime, Anywhere, Any Place. It was removed from the RNZAF register in 1948 and subsequently scrapped.

CHAPTER 27
Royal New Zealand Air Force

Nos. 40 and 41 Squadrons, RNZAF, operated C-47s throughout New Zealand and the South Pacific from 1943 onwards. All were C-47A and C-47B models which were delivered in the standard USAAF Olive Drab/grey colour scheme. The RNZAF roundel, later with a bar, was painted over the factory applied US insignia upon arrival in New Zealand. A total of 58 was delivered, and these were allocated RNZAF aircraft codes NZ3501 to NZ3558. These transports carried US and New Zealand servicemen and materiel throughout the SOPAC, and between the Solomons and New Zealand on leave, a route they shared with their USMC R4D counterparts.

Several were repainted in New Zealand after becoming worn in the tropics. These used available workshop paints, mostly locally manufactured green and Olive Drab colours from Balm Paints in Wellington. Several of these repaint schemes are indicated by a sharp wavy line between the over-painted Olive Drab and under surface grey.

The worst and only RNZAF C-47 loss in the South Pacific occurred in the immediate post-war period when NZ3526 disappeared during a flight from Espirito Santo to New Zealand on 25 September 1945 with twenty aboard. During this immediate post-war period No. 41 Squadron's NZ3534 was converted into an air ambulance for the repatriation of Allied POWs.

A line-up of No. 40 Squadron Dakotas and Lockheed Lodestars at Whenupai. The foremost aircraft is NZ3503, C-47A serial 42-23558.

RNZAF

95

96

Profile 95 – C-47A serial #42-23885, NZ3501, No. 40 Squadron, *Popeye III*

This C-47A was the first delivered to the RNZAF on 1 April 1943. However, the *Popeye III* nose art did not appear on the aircraft until sometime in 1945. The airframe was finally sold for scrap in 1962.

Profile 96 – C-47A serial #42-23556, NZ3502, No. 40 Squadron

Delivered to the RNZAF on 20 May 1943, this C-47A was repainted later in the war. The precise hue of the new green is unclear however it likely resembled Forest Green as illustrated here. The RNZAF followed the USAAF practice of adding bars to its insignia from September 1943.

C-47A NZ3502, the subject of Profile 96, is unloaded at Munda in late 1943.

The Popeye artwork on NZ3501, the subject of Profile 95, at Henderson Field in 1943. The name has not yet been painted next to the artwork.

A Transport Afdeeling crew at Archerfield in September 1943: note the Dutch epaulettes. VHCHK (serial number 42-23959) was first assigned into the 39th TCS as Hells Bells where it was often flown by Dutch crews. It was later lost to a forced-landing near Port Moresby on 14 September 1944. Note there was another Hells Bells (II) flown by the 41st TCS (see Profile 29).

CHAPTER 28
Dutch Transport Units

Following the surrender of the Netherlands East Indies in March 1942, a significant number of Dutch civilian and military transport aircraft were flown to Australia. These were modern Lockheed and Douglas types which formed the largest pool of transports in the country, and the aircraft were subsequently acquired by the USAAF where many entered service with the 22nd Transport Squadron (see Chapter 5).

The Dutch crews of these aircraft, some of them highly experienced, were called upon for military service. Many subsequently served with No. 18 (Netherlands East Indies) Squadron which was established in April 1942. It operated B-25s flown by Dutch crews but under RAAF administration.

The Dutch military crews had been part of the ML-KNIL (the Aviation Corps of the Netherlands East Indies Army), and in Australia these personnel were generally referred to as NEI Forces or the NEI Air Force (NEIAF). While large scale aircrew training for Dutch personnel was undertaken at Jackson Air Base in Mississippi from 1942-1944, some training also took place in Australia which enabled the gradual expansion of the NEIAF.

DAT was also keen on utilising experienced Dutch transport crews and accordingly held discussions with the provisional NEI government-in-exile in Melbourne. This led to the establishment of a Transport *Afdeeling* (TA) in Brisbane with 21 Dutch personnel assigned to the fledgling unit in December 1942.

The commanding officer of the TA (Brisbane) detachment was Major GA de Stoppelaar who also served as a liaison officer to the Fifth Air Force on all matters Dutch. The TA (Brisbane) was seconded to the Fifth Air Force and in January 1943 was established as a separate flight within the 39th TCS which had just arrived in Australia from the US (see Chapter 7).

The Dutch crews were quickly flying 39th TCS C-47s with the first flights being cargo runs to Port Moresby. It soon became apparent that the Dutch crews were far more experienced than their USAAF counterparts. The Americans had limited flight time in tropical conditions and the Dutch personnel were soon sharing their knowledge. The regular route initially flown was Brisbane-Townsville-Port Moresby delivering troops and materiel and returning with wounded soldiers or Japanese POWs.

Four TA (Brisbane) crews also participated in the Wau operation. Then from mid-February 1943 they were tasked with operating regular night services from Brisbane to Port Moresby carrying important administrative documents and also VIPs including General Douglas MacArthur, as they were considered the most capable pilots to do so. The courier/passenger flights became an exclusive TA (Brisbane) responsibility for the next one and a half years, a duty performed without a single incident. From September 1943 the Dutch crews also served with the 21st TCS.

From 26 August 1944 TA (Brisbane) was detached from the Fifth Air Force when the 21st TCS moved its operations to Nadzab. Meanwhile, a second Dutch unit, TA (Melbourne) had been created from Dutch crews based at RAAF Fairbairn, Canberra on 1 September 1943. It commenced operations with four C-60A Lodestars and one C-47A. These focused their movements within the Australian mainland mainly using the C-60As.

Meanwhile the need for air transport in the SWPA was increasing as Allied forces advanced, including into Dutch New Guinea. TA (Melbourne) subsequently became No. 1 Netherlands East Indies Transport Squadron (NEITS), while on 1 September 1944 TA (Brisbane) became No. 2 NEITS. Soon both units were regularly flying to new destinations in Dutch New Guinea, in support of the NEI government-in-exile (now based in Brisbane), as well as other Dutch military and civilian authorities. On 7 November 1944, all personnel from No. 2 NEITS were transferred to 1 NEITS, and on 15 August 1945 this combined unit became No. 19 (NEI) Squadron. The squadron developed a substantive transport capability and continued operating postwar. After transferring to Java in late 1946 the squadron was transferred to Dutch control at the start of 1947.

Allied Command had agreed in mid-1943 that a Dutch transport pool be established at Canberra. Initially eight C-47s were allocated, with the first being C-47A DT937 which arrived at Amberley near Brisbane on 22 December 1943. By the end of the war Dutch units were operating 34 C-47As in Australia. These aircraft and their Australian callsigns were:

DT-696 VHREW, DT-696 VHREW, DT-937 VHRDG, DT-938 VHRDH, DT-939 VHRDI, DT-940 VHROT, DT-941 VHRDK, DT-943 VHRDM, DT-944 VHRDN, DT-944 VHROM, DT-946 VHRDZ, DT-947 VHREA, DT-952 VHREF, DT-953 VHREG, DT-957 VHREK, DT-959 VHREM, DT-962 VHREP, DT-963 VHCJL, DT-963 VHREQ, DT-964 VHRER, DT-966 VHRET, DT-967 VHREU, DT-970 VHREX, DT-973 VHRCO, DT-975 VHRCQ, DT-977 VHRCR, DT-980 VHRCV, DT-981 VHRCZ, DT-985 VHRCB, DT-985 VHRCB, DT-986 VHRCD, DT-989 VHRCG, DT-990 VHRDH and DT-991 VHRCI.

C-47A DT-939, the subject of Profile 97, at Hollandia in mid-1944.

C-47A DT-944 Grobak, the subject of Profile 98, at Archerfield in October 1944.

C-47 DT-940 (serial number 42-92311) at Batchelor in the Northern Territory in Febuary 1945. It was no doubt on a support flight for No. 18 (NEI) Squadron which was based at Batchelor at that time.

NEI Transport Units

97

98

99

100

Profile 97 – C-47A serial #42-92474, DT-939, callsign VHRDI

This aircraft was received by the NEIAF as DT-939 on 17 February 1944. After the war it became Dutch civilian registration PK-RDI and then in April 1950 went to the Indonesian Air Force as T-439.

Profile 98 – C-47A serial #42-93434, DT-944, callsign VHRDN, *Grobak*

This aircraft was received by the NEIAF on 18 May 1944. After the war it flew with KLM then in June 1950 went to the Indonesian Air Force. The word *Grobak* is a Dutch East Indies colloquial word for a low-slung cart which carries heavy loads.

Profile 99 – C-47A serial #42-23584, callsign VHCGL, *Hairless Joe*

This transport commenced service in Australia with the 39th TCS on 1 June 1943, where it was operated exclusively by Dutch crews. On 28 October 1943 an NEI crew delivered engineers and parts including tyres to a stranded B-24D which had force-landed on salt flats some 30 miles from isolated Drysdale Mission in Australia's northwest. The crew had to deflate the C-47A's tyres and use wooden mats to facilitate departure, however, the aircraft became a complete loss when it ground-looped on take-off. Note the letter "L" behind the cockpit representing the last letter of the radio callsign.

Profile 100 – C-47A serial #42-93015, DT-941, callsign VHRDK

This aircraft was received by the NEIAF on 2 April 1944. It was written off at Cairns on 6 September 1944.

C-47A serial number 42-23584 Hairless Joe, the subject of Profile 99, seen in Brisbane prior to its loss in October 1943.

SOURCES & ACKNOWLEDGMENTS

Research for this volume draws exclusively from primary sources. It will surprise many that most records pertaining to USAAF transport operations are held in Australian archives, simply as they were so closely entwined with Australian operations including DAT. The author's extensive collection of C-47 photos, letters, logbook entries of C-47 operations and notes from field trips is cited throughout. These contain information garnered over many years too numerous to further credit, excluding the sources listed below.

Ongoing thanks to the website www.pacificwrecks.com and its diligent owner, Justin Taylan. Also to Peter Dunn's www.ozatwar.com and its detailed information on C-47 crashes and crash sites in Australia.

Sources include, but are not limited to;

Allied Air Force Intelligence Summaries (AWM)

Allied Translator and Interpreter Section (ATIS) Reports

ANGAU patrol officer reports of Allied crash sites, 1940s-1970s

Douglas Aircraft Corporation Historical Records

Cairns Historical Society

MacArthur Archives, MacArthur Memorial, Norfolk, VA

Field records and notes of Bill Chapman, Former Chemist in Port Moresby in the 1960s

Numerous Field Trips by the author within New Guinea and the Pacific, 1964-2017

Pacific Aircraft Historical Society – Wreck Data Sheets

Papua New Guinea Colonial Office - Civil Administration Records

Papua New Guinea Cultural Museum

RAAF Museum - Log Book Entries, Townsville Control Tower April 1942

USAAF Historical Study #17 – Air Action in the Papuan Campaign

Papua New Guinea Catholic Mission Association

Field Trips of John Douglas, Papua New Guinea, 1980s onwards

Field trips with James Luk, PNG 1976

Fifth and Thirteenth AF microfilms via Maxwell AFB (all relevant USAAF units cited in text)

History, 39th TCS, January 1943–January 1944.

Ford, My New Guinea Diary (memoirs of Staff Sergeant Ford, a pilot with the 6th TCS)

Kelly, Allied Air Transport Operations (primarily a reference volume citing unit records).

Atherton, "Flying Overseas." Philip Brinson, son of a 317th TCG veteran

History, 40th TCS, February 1942–January 1944

Headquarters Fifth Air Force Special Orders No. 272, 29 September 1943

Military records former KNILM/KLM personnel of NEI 18 Squadron

O.G. Ward, De Militaire Luchtvaart van het KNIL in de Jaren 1942-1945

Organisatie Transport Afdeeling Brisbane der Militaire Luchtvaart, No.2, Brisbane 1 Juli 1944

Report on conference dated 17 March 1942 between Lt-General Brett, Van Mook, Hoogstraten (director of

Economic Affairs), Colonel Giebel, Major Roos, Group Captain Harold Gatty and Colonel Perrin

USAFIA (Brett) telegrams to CG AAF in Washington about transfers of NEI and USAAF aircraft

Aircraft status reports 21st TS and 22nd TS early May 1942

RAAF Form E/E.88 (Status Cards) pertaining to C-47s and other Douglas Types

Aircraft Assigned to Air Transport Command since 26 January 1942, HQ, ATC

List of DC-3 aircraft operated by TAA, compiled by John Forsyth, Department of Transport

VH-registered Douglas DC-2/3/5 Aircraft, Part One – compiled by Al Bovelt 25 Apr 1998

Australian National Airways Pty Ltd, The Named Fleet, John Hopton

Australian Aeronautical Historical Record (Misc lists)

RAAF Aircraft Register 1951 to 1968, RAAF Museum, Point Cook, Victoria.

Historic Civil Aircraft Register of Australia, VH-CAA to VH-CZZ, Tony Arbon & David Sparrow

Back Load – Unofficial history of 433rd TCG

Two Years C/O Postmaster – Unofficial History of 13th TCS

History of 374th TCG – Edward Imparato

Dedication

Colonel Paul Prentiss at his operations desk at Nadzab, New Guinea, in 1944.

This volume is dedicated to Colonel Paul Prentiss, the commander of the 54th Troop Carrier Wing. Prentiss arrived in Australia in August 1942 just before the creation of the Fifth Air Force and was subsequently appointed Technical Inspector of Fifth Air Force Air Service Command. He then served as commander of the 374th Troop Carrier Group until May 1943, when he became commander of the 54th TCW, making him responsible for all USAAF transport operations in frontline New Guinea. In March 1944 he again headed Air Service Command, then in August 1945 he headed Far East Air Force Service Command.

Prentiss stayed on in the post-war USAF but on 20 July 1953, only a few weeks after he retired to his hometown of San Antonio, Texas, he was found deceased in his backyard. He had accidently electrocuted himself when conducting home renovations with an electric circular saw.

Index of Names

Adams, Major Fred 35

Adams, Lieutenant Richard 107

Anders, Lieutenant Philip 87

Anderson, Major Donald 93

Anderson, Lieutenant Colonel Joseph 12

Ashwell, Major Leonard 136

Baird, Major Erwin 103

Beaver, Lieutenant George 25

Beck, Colonel AJ 71

Beebe, Major Robert 37

Bell, Major Robert 136

Bergstrom, Lieutenant Don 29

Berry, Sergeant Ed 108

Beswick, Captain George 75

Billmaier, Corporal Lawrence 25

Bishop, Lieutenant Paul 93

Bonner, Major Joseph 107

Bradford, Major William 37

Brandt, Staff Sergeant Marvin 25

Brett, Major-General George 119

Bronson, Hubert 25

Calliham, Major Marvin 93

Chase, Lieutenant Gale 75

Church, Captain Jack 136

Cooper, Gary 69, 125

Crecelius, Lieutenant William 39

Davis, Lieutenant Marvin 79

de Stoppelaar, Major GA 153

Dobrisky, Lieutenant Andrew 103

Dunlop, Eddy 123

Eisenhower, General Dwight 9

Ellis, Lieutenant Henry 59

Elsmore, Colonel Ray 120

Evans, Captain James 65

Faught, Lieutenant Courtney 15

Feeney, Lieutenant Francis 37

Fink, Lieutenant Merrill 87

Fishburn, Lieutenant Mayhew 111

Fitch, Vice Admiral Aubrey 14

Flanigan, Major Robert 111

Fleps, Major Carl 136

Fonda, Henry 29

Ford, Lieutenant Joseph 47

Gatty, Group Captain Harold 11, 119

Gower, Second Lieutenant Lorenzo 67

Guess, Captain Vernon 93

Guillote, Major Dwight 136

Hampton, Major Edgar 35

Hardeman, Captain Milton 83

Head, Captain William 103

Healy, Lieutenant Robert 89

Henebry, Colonel "Jock" 13

Henry, Captain Fred 37

Hope, Bob 59

Hosp, Lieutenant Richard 76

Hutchinson, Lieutenant John 41, 93

Jackson, Captain Eugene 41

Jacobs, Lieutenant Paul 99

Jacques, Captain Pierre 37

James, Captain Leroy 136

Johnson, Major Harold 135

Johnston, Private Edward 51

Johnston, George 21

Keeler, Major Douglas 135

Kelley, Lieutenant Charles 76

King, Lieutenant Benjamin 79

King, Lieutenant Charles 3

Kobayashi Katsutauru, FPO1c 25

Lackey, Lieutenant Colonel John 15, 23

Lalonde, Sergeant Joseph 29

Lane, Major Henry 135

Lanman, Major William 136

Larson, Lieutenant JD 97

Lattier, Lieutenant Earl 25

Leahy, Richard 81

Leek, Lieutenant Colonel Frederick 136

Long, Lieutenant Malcolm 93

MacArthur, General Douglas 9, 15, 123, 153

Maruoka Yasuhei, Colonel 14

McClelland, Captain Hamish 23

McDonald, Major John 115

McGinnis, Lieutenant Thomas 116

McIllivray, Captain James 97

Merkling, Al 71

Moran, Major James 135

Morrison, Lieutenant Russell 115

Munsch, Lieutenant Colonel Albert 136

Nasset, Captain Erling 27

Nollkamper, First Lieutenant James 64

Okabe Toru, Major General 14

Patterson, Staff Sergeant Claude 44

Penn, Captain Perry 37

Peterson, Captain William 23

Pitts, Lieutenant Colonel Joel 15

Powell, Captain Lucian 79

Prentiss, Colonel Paul 15, 59, 159

Redfield, Major Ben 136

Richardson, Lieutenant Leonard 89

Rickenbacker, Captain Eddie 41

Sams, Lieutenant Robert 51

Sanford, Major Theodore 136

Schwensen, Lieutenant Robert 43

Seeds, Major Elmore 136

Smith, Major Donald 15, 97

Smith, Lieutenant Colonel Perry 135

Stanwyck, Barbara 29

Stenglein, Lieutenant Robert 21, 108

Sweetser, Major Warren 136

Swenson, Captain Raymond 37

Taylan, Justin 64

Teague, Flying Officer William 65

Van Liew, Major Harry 136

Vargas, Alberto 84

Wagoner, Staff Sergeant Kenneth 25

Waldman, Captain Herbert 57

Wamsley, Captain George 41

White, Captain Thomas 83

Whitehead, General Ennis 71

Williams, Major Bill 15

Williams, Major Freeman 135, 136

Woods, Major Donald 97

Yeaman, Captain Ralph 136